RECONCILING OBSERVATIONS OF GLOBAL TEMPERATURE CHANGE

Panel on Reconciling Temperature Observations
Climate Research Committee
Board on Atmospheric Sciences and Climate
Commission on Geosciences, Environment, and Resources

National Research Council

NATIONAL ACADEMY PRESS
Washington, D.C.

NOTICE: The project that is the subject of this report was approved by the Governing Board of the National Research Council, whose members are drawn from the councils of the National Academy of Sciences, the National Academy of Engineering, and the Institute of Medicine. The members of the committee responsible for the report were chosen for their special competences and with regard for appropriate balance.

Support for this project was provided by the National Oceanic and Atmospheric Administration under Contract No. 50-DKNA-7-90052 and by Alcoa. Any opinions, findings, and conclusions or recommendations expressed in this publication are those of the author(s) and do not necessarily reflect the views of the NOAA or any of its sub-agencies or of Alcoa.

International Standard Book Number 0-309-06891-6

Additional copies of this report are available from:

National Academy Press
2101 Constitution Avenue, NW
Box 285
Washington, D.C. 20055
800-624-6242
202-334-3313 (in the Washington Metropolitan Area)
www.nap.edu

Cover: Surface and lower to mid-tropospheric temperature trends for the period 1979-1998. The surface data (left panel) are comprised of surface air temperature over land and the temperature of water at the ocean's surface, and have been subjected to a slight additional smoothing to simplify the pattern (Jones et al., 1999). The lower to mid-tropospheric data (right panel) are derived from satellite observations from the Microwave Sounding Unit Channel 2 (the so-called "MSU 2LT") (Christy et al., 2000). For both datasets, the trends are computed using the method of ordinary least squares. The color key is the same as in Figure 6.2. The map views on the front cover are centered at 30° N and 110° W and the views on the back cover are centered at 30° S and 70° E. For the globe as a whole (see Figures 6.2 and 7.1 inside), warming has been prevalent at the earth's surface, but much less so in the lower to mid-troposphere.

Printed in the United States of America

THE NATIONAL ACADEMIES

National Academy of Sciences
National Academy of Engineering
Institute of Medicine
National Research Council

The **National Academy of Sciences** is a private, nonprofit, self-perpetuating society of distinguished scholars engaged in scientific and engineering research, dedicated to the furtherance of science and technology and to their use for the general welfare. Upon the authority of the charter granted to it by the Congress in 1863, the Academy has a mandate that requires it to advise the federal government on scientific and technical matters. Dr. Bruce M. Alberts is president of the National Academy of Sciences.

The **National Academy of Engineering** was established in 1964, under the charter of the National Academy of Sciences, as a parallel organization of outstanding engineers. It is autonomous in its administration and in the selection of its members, sharing with the National Academy of Sciences the responsibility for advising the federal government. The National Academy of Engineering also sponsors engineering programs aimed at meeting national needs, encourages education and research, and recognizes the superior achievements of engineers. Dr. William A. Wulf is president of the National Academy of Engineering.

The **Institute of Medicine** was established in 1970 by the National Academy of Sciences to secure the services of eminent members of appropriate professions in the examination of policy matters pertaining to the health of the public. The Institute acts under the responsibility given to the National Academy of Sciences by its congressional charter to be an adviser to the federal government and, upon its own initiative, to identify issues of medical care, research, and education. Dr. Kenneth I. Shine is president of the Institute of Medicine.

The **National Research Council** was organized by the National Academy of Sciences in 1916 to associate the broad community of science and technology with the Academy's purposes of furthering knowledge and advising the federal government. Functioning in accordance with general policies determined by the Academy, the Council has become the principal operating agency of both the National Academy of Sciences and the National Academy of Engineering in providing services to the government, the public, and the scientific and engineering communities. The Council is administered jointly by both Academies and the Institute of Medicine. Dr. Bruce M. Alberts and Dr. William A. Wulf are chairman and vice chairman, respectively, of the National Research Council.

PANEL ON RECONCILING TEMPERATURE OBSERVATIONS

Members

JOHN M. WALLACE (*Chair*), University of Washington, Seattle
JOHN R. CHRISTY, University of Alabama in Huntsville
DIAN J. GAFFEN, NOAA/Air Resources Laboratory, Silver Spring, Maryland
NORMAN C. GRODY, NOAA/NESDIS, Camp Springs, Maryland
JAMES E. HANSEN, NASA Goddard Institute for Space Studies, New York, New York
DAVID E. PARKER, Hadley Centre, Meteorological Office, Bracknell, United Kingdom
THOMAS C. PETERSON, NOAA/National Climatic Data Center, Asheville, North Carolina
BENJAMIN D. SANTER, Lawrence Livermore National Laboratory, Livermore, California
ROY W. SPENCER, NASA Marshall Space Flight Center, Huntsville, Alabama
KEVIN E. TRENBERTH, National Center for Atmospheric Research, Boulder, Colorado
FRANK J. WENTZ, Remote Sensing Systems, Santa Rosa, California

Consultant

TODD MITCHELL, University of Washington, Seattle

NRC Staff

PETER A. SCHULTZ, Study Director
DIANE L. GUSTAFSON, Administrative Assistant

CLIMATE RESEARCH COMMITTEE

Members

EUGENE M. RASMUSSON (*Chair*), University of Maryland, College Park
EDWARD S. SARACHIK (*Vice-Chair*), University of Washington, Seattle
MAURICE BLACKMON, National Center for Atmospheric Research, Boulder, Colorado
JEFF DOZIER, University of California, Santa Barbara
JAMES GIRAYTYS, Consultant, Winchester, Virginia
JAMES E. HANSEN, NASA Goddard Institute for Space Studies, New York, New York
PHILIP E. MERILEES, Naval Research Laboratory, Monterey, California
ROBERTA BALSTAD MILLER, CIESIN, Columbia University, Palisades, New York
S. ICHTIAQUE RASOOL, International Consultant, Paris, France
STEVEN W. RUNNING, University of Montana, Missoula
ANNE M. THOMPSON, NASA Goddard Space Flight Center, Greenbelt, Maryland
ANDREW WEAVER, University of Victoria, British Columbia
ERIC F. WOOD, Princeton University, New Jersey

Ex Officio Members

W. LAWRENCE GATES, Lawrence Livermore National Laboratory, Livermore, California
DOUGLAS G. MARTINSON, Lamont-Doherty Earth Observatory of Columbia University, Palisades, New York
JOHN O. ROADS, Scripps Institution of Oceanography, La Jolla, California

NRC Staff

PETER A. SCHULTZ, Program Director
CARTER W. FORD, Project Assistant

BOARD ON ATMOSPHERIC SCIENCES AND CLIMATE

Members

ERIC J. BARRON (*Co-Chair*), Pennsylvania State University, University Park
JAMES R. MAHONEY (*Co-Chair*), Consultant, McLean, Virginia
SUSAN K. AVERY, Cooperative Institute for Research in Environmental Sciences, University of Colorado, Boulder
LANCE F. BOSART, State University of New York, Albany
MARVIN A. GELLER, State University of New York, Stony Brook
CHARLES E. KOLB, Aerodyne Research, Inc., Billerica, Massachusetts
ROGER A. PIELKE, JR., National Center for Atmospheric Research, Boulder, Colorado
ROBERT T. RYAN, WRC-TV, Washington, D.C.
MARK R. SCHOEBERL, NASA Goddard Space Flight Center, Greenbelt, Maryland
JOANNE SIMPSON, NASA Goddard Space Flight Center, Greenbelt, Maryland
NIEN DAK SZE, Atmospheric and Environmental Research, Inc., Cambridge, Massachusetts
ROBERT A. WELLER, Woods Hole Oceanographic Institution, Massachusetts
ERIC F. WOOD, Princeton University, New Jersey

Ex Officio Members

DONALD S. BURKE, Johns Hopkins University, Baltimore, Maryland
DARA ENTEKHABI, Massachusetts Institute of Technology, Cambridge
MICHAEL C. KELLEY, Cornell University, Ithaca, New York
JOHN O. ROADS, Scripps Institution of Oceanography, La Jolla, California
EUGENE M. RASMUSSON, University of Maryland, College Park
PAUL WINE, Georgia Institute of Technology, Atlanta

NRC Staff

ELBERT W. (JOE) FRIDAY, JR., Director
LAURIE S. GELLER, Program Officer

PETER A. SCHULTZ, Program Officer
DIANE L. GUSTAFSON, Administrative Assistant
ROBIN MORRIS, Financial Associate
TENECIA A. BROWN, Senior Program Assistant
CARTER W. FORD, Project Assistant

COMMISSION ON GEOSCIENCES, ENVIRONMENT, AND RESOURCES

Members

GEORGE M. HORNBERGER (*Chair*), University of Virginia, Charlottesville
RICHARD A. CONWAY, Union Carbide Corporation (Retired), S. Charleston, West Virginia
THOMAS E. GRAEDEL, Yale University, New Haven, Connecticut
THOMAS J. GRAFF, Environmental Defense Fund, Oakland, California
EUGENIA KALNAY, University of Maryland, College Park
DEBRA KNOPMAN, Progressive Policy Institute, Washington, D.C.
KAI N. LEE, Williams College, Williamstown, Massachusetts
BRAD MOONEY, J. Brad Mooney Associates, Ltd., Arlington, Virginia
HUGH C. MORRIS, El Dorado Gold Corporation, Vancouver, British Columbia
H. RONALD PULLIAM, University of Georgia, Athens
MILTON RUSSELL, Joint Institute for Energy and Environment and University of Tennessee (Emeritus), Knoxville
THOMAS C. SCHELLING, University of Maryland, College Park
ANDREW R. SOLOW, Woods Hole Oceanographic Institution, Woods Hole, Massachusetts
VICTORIA J. TSCHINKEL, Landers and Parsons, Tallahassee, Florida
E-AN ZEN, University of Maryland, College Park
MARY LOU ZOBACK, U.S. Geological Survey, Menlo Park, California

NRC Staff

ROBERT M. HAMILTON, Executive Director
GREGORY H. SYMMES, Associate Executive Director
JEANETTE SPOON, Administrative and Financial Officer
DAVID FEARY, Scientific Reports Officer
SANDI FITZPATRICK, Administrative Associate
MARQUITA SMITH, Administrative Assistant/Technology Analyst

Acknowledgments

This report has been reviewed in draft form by individuals chosen for their diverse perspectives and technical expertise, in accordance with procedures approved by the NRC's Report Review Committee. The purpose of this independent review is to provide candid and critical comments that will assist the institution in making the published report as sound as possible and to ensure that the report meets institutional standards for objectivity, evidence, and responsiveness to the study charge. The review comments and draft manuscript remain confidential to protect the integrity of the deliberative process. We thank the following individuals for their participation in the review of this report:

JAMES ANGELL, NOAA/Air Resources Laboratory
ALAN BASIST, NOAA/National Climatic Data Center
LENNART BENGTSSON, Max Planck Institute for Meteorology
SIMON BROWN, Hadley Centre, Meteorological Office, United Kingdom
JAMES HOLTON, University of Washington
JAMES HURRELL, National Center for Atmospheric Research
EUGENIA KALNAY, University of Maryland
RICHARD LINDZEN, Massachusetts Institute of Technology
NEVILLE NICHOLLS, Australian Bureau of Meteorology Research Centre
EUGENE M. RASMUSSON, University of Maryland

While the individuals listed above have provided constructive comments and suggestions, it must be emphasized that responsibility for the final content of this report rests entirely with the authoring committee and the institution.

The panel wishes to thank Todd Mitchell at the University of Washington for his contributions and insight in the presentation of the report's complex data, Jay Lawrimore at the National Climatic Data Center for supplying figures and data, and David Feary at the National Research Council for his editorial guidance.

Preface

A National Research Council panel was convened to examine observed trends of temperature near the surface and in the lower to mid-troposphere (the atmospheric layer extending from the earth's surface up to about 8 km). The objectives of this panel were to:

(1) summarize the state of the science in the measurement of temperature from space, from radiosondes, and from surface instrumentation;
(2) assess the biases and uncertainties in the data;
(3) describe the major conflicts in the trends; and
(4) define the actions required to reduce the uncertainties and biases.

The panel, which is under the purview of the Board on Atmospheric Sciences and Climate's (BASC) Climate Research Committee (CRC), included individuals with expertise on all relevant technical facets of the issue.

The panel's report, presented here, is structured in a layered fashion, providing the reader with an increasing level of technical detail. The Executive Summary gives a very brief overview of the report's findings and recommendations and is targeted towards non-scientists. Part I of the main body of the report is intended for the public, policy-making, and scientific communities and is also written in a relatively non-technical fashion. Part I includes a chapter outlining the key questions (Introduction) and another that provides an overview of the relevant measurement types and their observations (Background). Part I concludes with chapters on the panel's Findings and Recommendations. Part II more fully articulates the scientific basis for the discussion and conclusions that are presented in Part I, by detailing the major, relevant measurement systems and their temperature records. It does so in chapters on Surface Temperature Observations, MSU Observations, and Radiosonde Observations. Part II concludes with a chapter that compares the temperature records of the three types of observations and presents possible reasons for the observed temperature trend differences.

Contents

Executive Summary

The global-mean temperature at the earth's surface is estimated to have risen by 0.25 to 0.4 °C during the past 20 years. On the other hand, satellite measurements of radiances indicate that the temperature of the lower to mid-troposphere (the atmospheric layer extending from the earth's surface up to about 8 km) has exhibited a smaller rise of approximately 0.0 to 0.2 °C during this period. Estimates of the temperature trends of the same atmospheric layer based on balloon-borne observations (i.e., radiosondes) tend to agree with those inferred from the satellite observations. The panel was asked to assess whether these apparently conflicting surface and upper air temperature trends lie within the range of uncertainty inherent in the measurements and, if they are judged to lie outside that range, to identify the most probable reason(s) for the differences.

To address these questions the panel had to consider:

- the factors that contribute to uncertainties in the trends inferred from three categories of instrumental measurements—Microwave Sounding Units (MSU) carried aboard National Oceanic and Atmospheric Administration (NOAA) satellites, radiosondes, and surface observations;
- the technical issues involved in making comparisons between global-mean temperature trends derived from measurements with different

physical characteristics, different spatial and temporal sampling characteristics, and different error characteristics;

- the impact of the recent corrections to the algorithms for processing measurements derived from the MSU to account for satellite drifting and changes in instrument response;
- the contribution of natural climate variability to decade-to-decade climate changes, including changes in the atmosphere's vertical structure associated with natural variability;
- the changes in the atmosphere's vertical structure associated with human-induced climate changes; and
- the results of recent climate model simulations of temperature trends that take into account both natural variability and human-induced forcing.[1]

In the opinion of the panel, the warming trend in global-mean surface temperature observations during the past 20 years is undoubtedly real and is substantially greater than the average rate of warming during the twentieth century. The disparity between surface and upper air trends in no way invalidates the conclusion that surface temperature has been rising. The recent corrections in the MSU processing algorithms (referred to above) bring the global temperature trend derived from the satellite data into slightly closer alignment with surface temperature trends, but a substantial disparity remains. The various kinds of evidence examined by the panel suggest that the troposphere actually may have warmed much less rapidly than the surface from 1979 into the late 1990s, due both to natural causes (e.g., the sequence of volcanic eruptions that occurred within this particular 20-year period) and human activities (e.g., the cooling of the upper part of the troposphere resulting from ozone depletion in the stratosphere). Regardless of whether the disparity is real, the panel cautions that temperature trends based on data for such short periods of record, with arbitrary start and end points, are not necessarily indicative of the long-term behavior of the climate system.

Reducing uncertainties in the evaluation of the trends will require: (1) implementing an improved climate monitoring system designed to ensure the continuity and quality of critically needed measurements of

---

[1] A climate forcing is a perturbation to the energy balance of the earth–atmosphere system and may bring about climate change.

temperature, other climatic variables, and concentrations of aerosols and trace gases; and (2) making raw and processed atmospheric measurements accessible in a form that enables a number of different groups to replicate and experiment with the processing of the more widely disseminated data sets such as the MSU tropospheric temperature record. A number of possible research strategies for improving the understanding of uncertainties inherent in the various measurement systems and the relationship between surface and upper air temperature trends are proposed in the report.

Part I

Overview and Conclusions

1

Introduction

An overall increase in global-mean atmospheric temperatures is predicted to occur in response to human-induced increases in atmospheric concentrations of heat-trapping "greenhouse gases" (IPCC, 1996). The most prominent of these gases, carbon dioxide, has increased in concentration by over 30% during the past 200 years, and is expected to continue to increase well into the future. Other changes in atmospheric composition complicate the picture. In particular, increases in the number of small particles (called aerosols) in the atmosphere regionally offset and mask the greenhouse effect, and stratospheric ozone depletion contributes to cooling of the upper troposphere and stratosphere.[2,3]

Many in the scientific community believe that a distinctive greenhouse-warming signature is evident in surface temperature data for the past few decades. Some, however, are puzzled by the fact that satellite temperature measurements indicate little, if any, warming of the lower to mid-troposphere (the layer extending from the surface up to about 8 km) since such satellite observations first became operational in

[2] The troposphere is the atmospheric layer where the temperature generally decreases with height, extending from the surface up to approximately 10-15 km, and the stratosphere is the stable layer above that extending up to approximately 50 km.

[3] Further complicating the response of the different atmospheric levels to increases in greenhouse gases are other processes such as those associated with changes in the concentration and distribution of atmospheric water vapor and clouds.

1979. The satellite measurements appear to be substantiated by independent trend estimates for this period based on radiosonde data. Some have interpreted this apparent discrepancy between surface and upper air observations as casting doubt on the overall reliability of the surface temperature record,[4] whereas others have concluded that the satellite data (or the algorithms that are being used to convert them into temperatures) must be erroneous. It is also conceivable that temperatures at the earth's surface and aloft have not tracked each other perfectly because they have responded differently to natural and/or human-induced climate forcing during this particular 20-year period. Whether these differing temperature trends can be reconciled has implications for assessing:

- how much the earth has warmed during the past few decades,
- whether observed changes are in accord with the predicted response to the buildup of greenhouse gases in the atmosphere based on model simulations, and
- whether the existing atmospheric observing system is adequate for the purposes of monitoring global-mean temperature.

This report reassesses the apparent differences between the temperature changes recorded by satellites and the surface thermometer network on the basis of the latest available information. It also offers an informed opinion as to how the different temperature records should be interpreted, and recommends actions designed to reduce the remaining uncertainties in these measurements.

[4] Unless specified otherwise, the "surface record" referred to in this report is a combination of the temperature of sea surface water and the temperature of surface air over land.

2

Background

Estimates of temperature variations near the earth's surface are based on thermometer readings taken daily at thousands of land stations and on board thousands of ships. Dating back into the late nineteenth century, the data coverage has been dense enough to reveal the existence of gradual changes in hemispheric- and global-mean surface temperature. A time series of global-mean temperature from 1880 to 1998 (Figure 2.1) displays short-term fluctuations that can be identified with El Niño events and volcanic eruptions. Superimposed upon these short-term fluctuations in the time series are more gradual variations that include a warming of between 0.4 and 0.8 °C over the course of the century. The exact amount of estimated warming depends upon which of the existing compilations of the data is used as a basis for the calculation, the method used to estimate global means on the basis of irregularly spaced station observations, and the way in which the data are smoothed in time. Such globally averaged time series are not necessarily representative of local conditions: for example, Canada and Siberia have warmed much more rapidly during the past 20 years than indicated in Figure 2.1, while parts of the high latitude North Atlantic and North Pacific regions have cooled slightly. In order to estimate globally averaged temperature changes with a high degree of accuracy, it is necessary to have a broad spatial distribution of observations that are made with high precision.

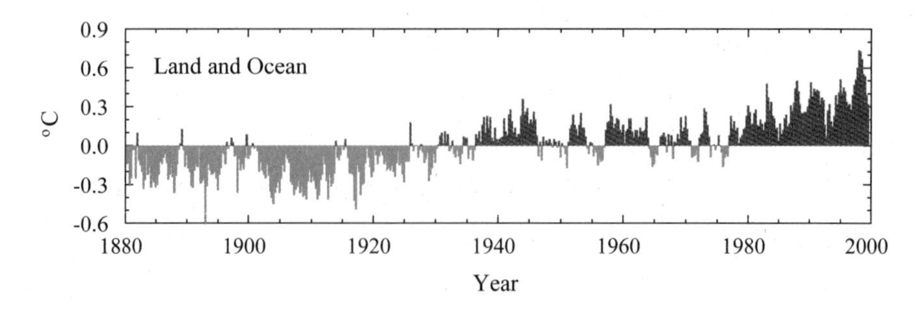

FIGURE 2.1. Time series of seasonally averaged global surface temperature (December 1879–August 1999) based on the Quayle et al. (1999) data set, computed as differences from the 1880–1998 mean. The time series uses an area-weighted average of the surface air temperature over land and the temperature of water at the ocean's surface.

Temperature changes at and just above the earth's surface are of singular importance from the standpoint of societal and human impacts, and they are also widely regarded as an important indicator of human-induced climate change. However, if global warming is caused by the build-up of greenhouse gases in the atmosphere, it should be evident not only at the earth's surface, but also in the lower to mid-troposphere. Temperatures aloft can be measured in a number of ways, two of which are useful for climate monitoring: by radiosondes (balloon-borne instrument packages, including thermometers, released daily or twice daily at a network of observing stations throughout the world), and by satellite measurements of microwave radiation emitted by oxygen gas in the lower to mid-troposphere, taken with an instrument known as the Microwave Sounding Unit (MSU).[5] The balloon measurements are taken at the same Greenwich mean times each day, whereas the times of day of the satellite measurements for a given location drift slowly with changes in the satellite orbits. The radiosonde network has been operative since the late 1940s and substantially increased its coverage during the International Geophysical Year (1957–58). However, it is

---

[5] The Microwave Sounding Unit senses radiation in a number of different channels, each of which is representative of a different layer of the atmosphere. The measurements discussed in this report are derived from channel 2—a channel that senses radiation in the layer extending from the surface up to about 15 km. To eliminate the influence of the stratospheric radiation, rather elaborate processing is required. The processed data are referred to as MSU 2LT (lower to mid-troposphere). Successive, improved versions of the MSU 2LT data have been produced over the past several years. The current version (D) was released in early 1999. For further discussion see chapter 7.

only since the mid-1960s that the instrumentation has been stable enough and sufficiently well documented for these measurements to be of use for estimating global temperature changes. Continuous MSU measurements began in 1979.

In the scientific literature on the detection of climate change, temperatures are commonly expressed in terms of departures from the local "climate" mean for a specified reference period. In this report, the reference period is the 20-year period 1979–98. Such departures from climate means are referred to as "temperature anomalies." Figure 2.2 shows the patterns of tropospheric temperature anomalies over the Western Hemisphere, as sensed by the MSU, during the northern winters (December through February) of 1982–83 and 1997–98, which both correspond to strong El Niño events in the tropical Pacific, and during the winter of 1988–89, which corresponds to a La Niña event. During the El Niño winters, temperatures throughout the tropics were above the mean of the past 20 years (i.e., the anomalies were positive), with alternating patches of warm and cold anomalies at higher latitudes. In contrast, during the La Niña winter the tropics were colder than the mean for the 20-year period. The patterns in this figure reflect the warmer global-mean temperatures characteristic of El Niño years, in contrast to the cooler La Niña years.

Figure 2.3 shows three time series of global-mean temperature anomalies. The black curve represents surface temperature, and the colored curves represent the temperature of the lower to mid-troposphere as inferred from MSU measurements (red) and radiosonde observations (green). Year-to-year fluctuations are evident in all three time series, and particularly in the series for the temperatures aloft. For example, the El Niño years 1983 and 1998 were a few tenths of a degree warmer, while 1992–93 following the eruption of Mt. Pinatubo were a few tenths of a degree cooler, than the 20-year average. Contrasting warm El Niño and cold La Niña years show up even more clearly in the tropical time series shown in Figure 2.4. In both global and tropical data, the peaks and dips in the satellite and radiosonde time series correlate quite well. Since these two time series represent largely independent mean temperature estimates for the same atmospheric layer, the strong correspondence between them is further proof that the fluctuations are real. El Niño and La Niña years are also evident in surface observations for the tropical belt (Figure 2.4), but they do not show up as clearly in the global-mean time series (Figure 2.3).

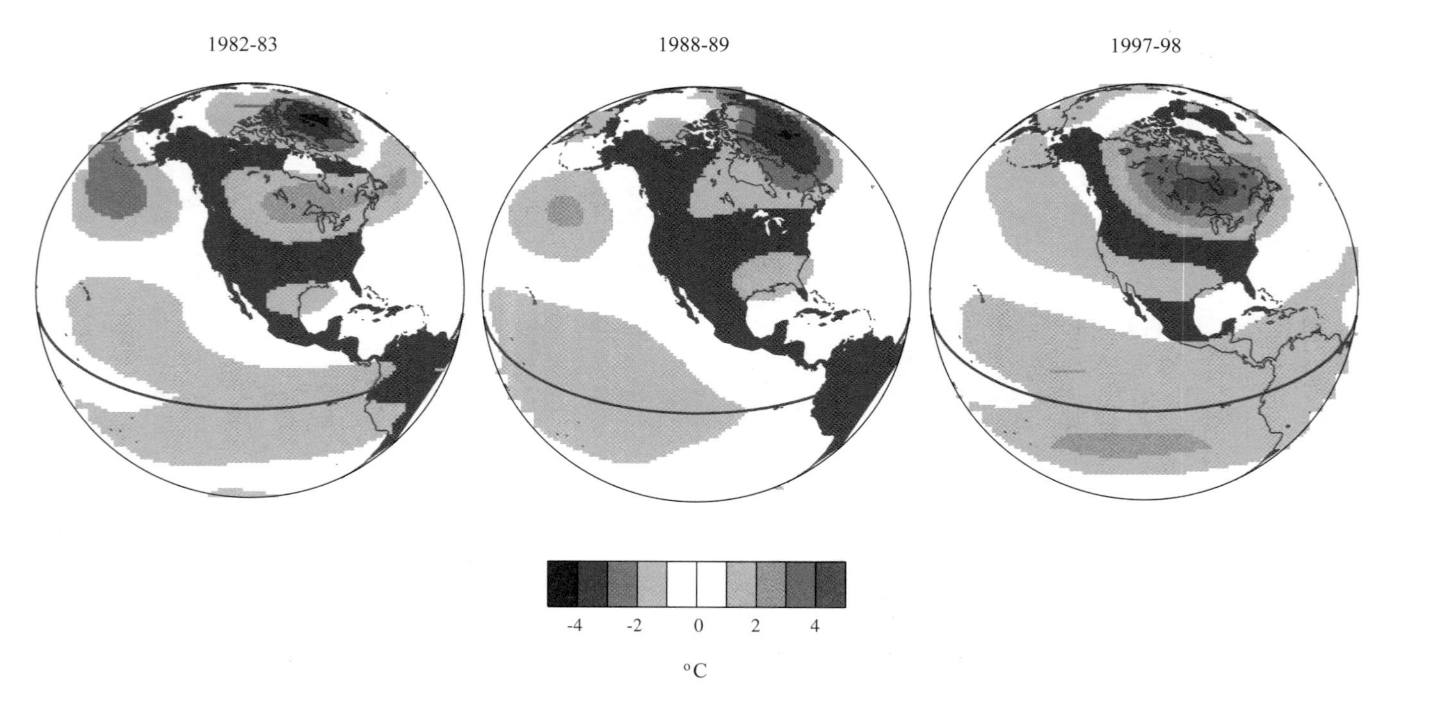

Figure 2.2. Upper air temperature anomalies over North America and the eastern Pacific Ocean for three different winters, computed as the average temperature during December, January, and February. Two of these winters were characterized by strong El Niños (1982–83 and 1997–98) and one by a strong La Niña (1988–89). The data are derived from the MSU Channel 2 and represent lower to mid-tropospheric temperature (the so-called “MSU 2LT”). The contour interval is 1 °C. Warm anomalies are indicated by orange/red tones and cold anomalies by blue tones.

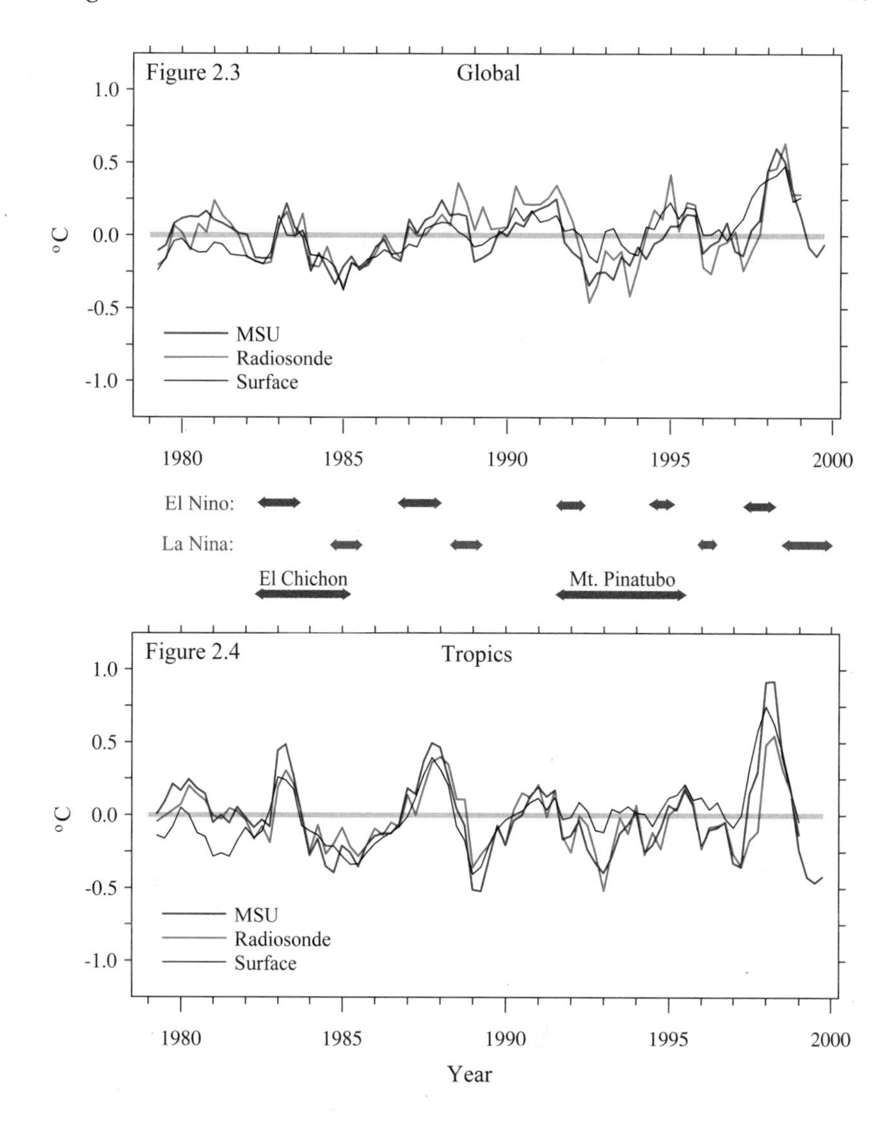

FIGURE 2.3. Seasonal mean time series of global-mean temperature anomalies from 1979 to 1998. The red curve shows lower to mid-tropospheric temperature from MSU 2LT (Christy et al., 2000). The green curve shows temperature data from the same layer as measured by radiosondes (i.e., "simulated 2LT") (Parker et al., 1997). The black curve shows surface temperature data (a combination of land air and sea based on the Jones et al. (1999) data set). The light gray line represents the mean of each time series. The first season is March–May 1979, and the last season is September–November 1999 for the MSU data set and December 1998–February 1999 for the other two data sets. El Niño, La Niña, and volcanic cooling episodes are indicated below the figure.

FIGURE 2.4. As in Figure 2.3, but for the tropical domain 20 °S – 20 °N.

Upon close inspection, it is evident that the surface temperature time series in both Figures 2.3 and 2.4 show upward trends relative to the corresponding tropospheric temperature time series for the past 20 years. The fit of a trend line to the time series of global-mean surface temperature (e.g., Figure 2.5) indicates a warming between 0.25 to 0.4 °C for this 20-year period, or approximately 0.1 to 0.2 °C per decade,[6] depending upon which of the existing data sets is used to represent the surface temperatures, and exactly how the fitting is done. In contrast, the tropospheric time series exhibits a smaller upward temperature trend of about 0.1 °C during this 20-year period. This disparity between the recent trends in global-mean surface and tropospheric temperature is the motivation for this report. Since this phenomenon first became apparent in the early 1990s, the research community has been seeking to identify and quantify possible sources of errors in the surface and upper air temperature measurements, and it has been trying to understand the physical processes that may have caused surface and upper air temperatures to change relative to one another. A number of biases in the data sets have been identified and corrected, and the process of refining the data sets is continuing.

In considering possible sources of errors in the satellite, radiosonde, and surface-based temperature measurements, it should be noted at the outset that none of these measurement systems was specifically designed for long-term climate monitoring (NRC, 1999). Changes in instrumentation and station locations have introduced time-varying biases into all three temperature time series. In principle, time series can be adjusted to remove these artifacts, but in practice there is some ambiguity in making such corrections. Decisions concerning which corrections need to be made, and how to implement them, are subject to debate. While many adjustments have been implemented, some quite recently, there will always remain a possibility of biases in the data that may be beyond the range of the current formal error estimates based on currently recognized sources of error. One mitigating factor is the

[6] In the literature on climate change, rates of change observed during prescribed intervals such as the past 20 years are conventionally expressed in units of degrees per decade. Rates of change computed in this manner are not necessarily applicable to periods of record outside the interval for which they were estimated. For example, the rate of warming of surface air temperature observed during the past 20 years is much greater than that observed during the previous 20-year interval, 1960-79, and is not necessarily indicative of the rate of temperature change that will be observed during the future interval 2000-2019.

independence of both the measurement errors and the uncertainties in satellite, radiosonde, and surface-based temperature records, which lends greater confidence to an assessment based on all three measurement categories than to an assessment based on any one of them in isolation.

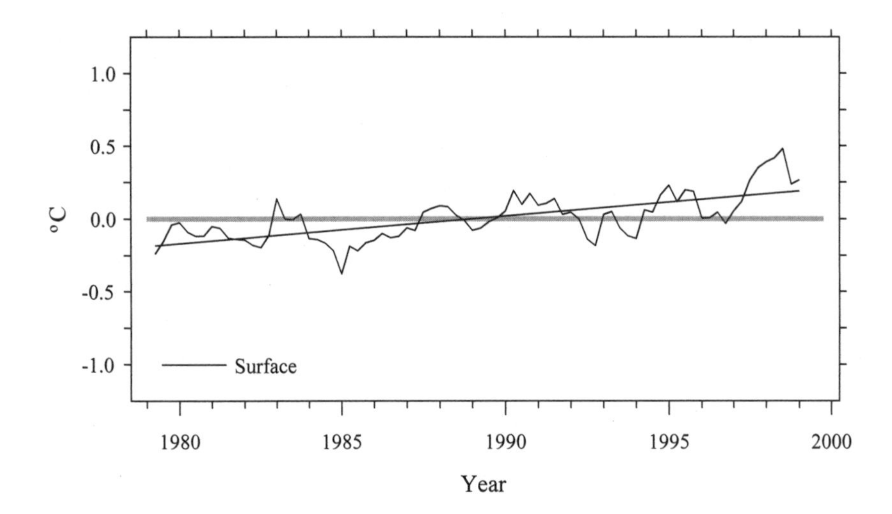

FIGURE 2.5. Time series of global-mean surface temperature from 1979 to 1998, repeated from Figure 2.3, shown with a trend line fitted by the method of ordinary least squares. The numerical value of the 20-year trend based on this particular data set (Jones et al., 1999) and fitting method is 0.19 °C/decade.

A concern that has been raised with respect to the surface-based temperature measurements is the effect of land use changes such as urbanization. As growing metropolitan areas encroach into the surroundings of formerly rural observing stations, the temperatures at these stations rise, particularly at night, in response to the well-documented "urban heat island effect." Some have suggested that much of the observed rise in global surface temperature during the twentieth century might be merely the expression of such local environmental transformations that are real, but not necessarily a signature of the global warming predicted to be associated with an increase in atmospheric greenhouse gas concentrations. These concerns have been addressed in numerous studies over the years that have sought to quantify the effect of land use changes and adjust the estimated global surface temperature trends accordingly. There have also been continuing efforts to document changes in instrumentation and observing practices, and to make appropriate adjustments in the data to compensate for them. Documentation of instrumentation and observing practices is also critical with respect to the radiosonde data. Ongoing efforts are being made to recover information on the past observing practices of the various national weather services and to apply adjustments as appropriate.

The major uncertainties in satellite measurements of upper air temperature are due to sensor and spacecraft biases and instabilities, the characteristics of which need to be estimated by performing satellite intercalibrations during overlapping intervals. These intervals are designed to be about two years long, but on two occasions, the overlap was substantially shorter due to instrument failures. The temperature measurements have recently been adjusted for gradual changes in satellite orbits that affect the levels and times of day at which the microwave radiation is sampled, and for small non-linearities in sensor performance, which cannot be determined in advance on the basis of laboratory calibrations. Because there is, in effect, only one satellite-based temperature record for which most of the processing has been performed by a single group, efforts to independently verify the MSU temperature measurements have, of necessity, focused on comparisons with radiosonde data.

Calculating the global-mean temperature anomaly for a particular season based on the MSU is straightforward, because the measurements are densely spaced and global in extent. However, for radiosonde observations, which are irregularly spaced with large gaps over the oceans (Figure 2.6), global-mean temperature is estimated on the basis of those stations operating during the season in question. Notice, for example, how the radiosonde data fail to sample the strongest local temperature anomalies over the subtropical eastern Pacific shown in Figure 2.2. Even in the absence of any real temperature variation, the global-mean temperature anomaly computed from radiosonde data could conceivably change from one season or decade to the next, merely as a result of stations in one of these poorly sampled regions going into or out of operation. Surface-based estimates are also subject to similar discontinuities, but they are not considered as serious a problem because there are so many more surface stations than there are radiosonde stations (compare Figures 2.6 and 2.7). In addition, surface data coverage over the oceans is much better, with the notable exception of high latitudes in the Southern Hemisphere. The effects of this uneven sampling are being investigated and quantified in several ways, for example by estimating “true” global-mean temperatures from the complete fields generated by satellite observations, blends of satellite and in situ data, or climate models, and then sampling these fields using the actual (incomplete) observed data coverage (see chapter 9).

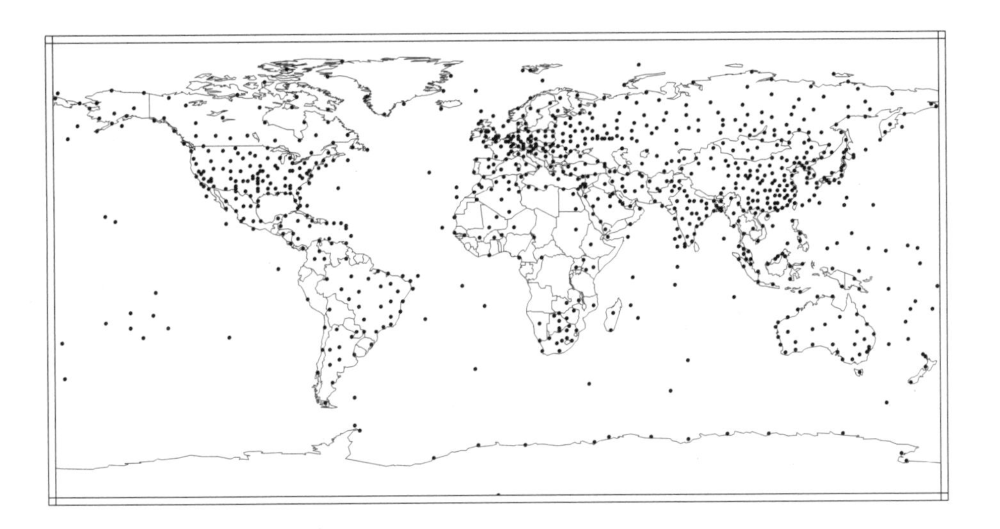

FIGURE 2.6. Location of radiosonde stations in the World Meteorological Organization's (WMO) Global Observing System upper air network. Fewer than half of these 905 stations report monthly climatic data, and only about two-thirds have reported regularly over the past few decades. Although most report some daily data, tropical portions of South America and Africa are often missing. Note the sparseness of the stations over the oceans and the high latitudes.

Measurement errors and uncertainties are not the whole story. The possibility that there may have been a real disparity between trends in surface and tropospheric temperature also needs to be considered. One way of making such an assessment is to consider whether simulations of the evolving climate of the past 20 years in climate models exhibit disparities as large as those that are observed.

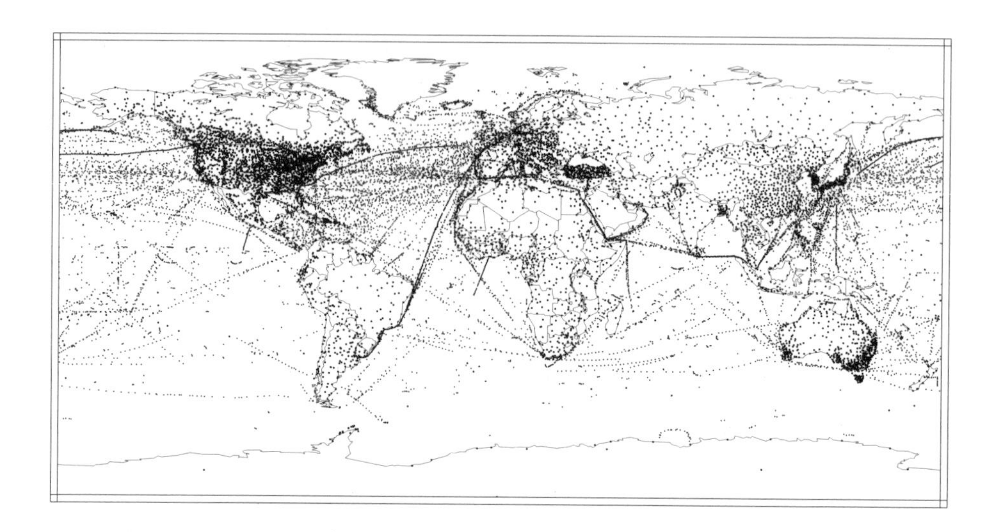

FIGURE 2.7. Location of surface temperature observations. The map is a composite of all of the Global Historical Climatology Network temperature stations (small dots), and the ship-based and floating and moored buoy observations from a single week used in the production of the Reynolds and Smith (1994) sea surface temperature data set (very small dots).

Climate models are tools that can be used to relate changes at the surface to those in the troposphere. Although today's state-of-the-art models accurately depict many physical processes, they are deficient in several respects, owing to difficulties in representing small-scale processes, such as those associated with clouds. Moreover, the detailed three-dimensional spatial structure and the temporal evolution of the many forcings of the climate system that are used to "drive" the models are poorly known. Model simulations are helpful in understanding the disparity between the 20-year trends in surface versus tropospheric temperatures, but they are not sufficiently reliable to provide a definitive assessment of whether the trends at these two levels are physically consistent.

Due to the non-deterministic nature of the climate system, an ensemble[7] of simulations run with the same climate model yields a

[7] In such an ensemble, each individual simulation is run with the same time-dependent climate forcings (greenhouse gases, aerosols, etc), but with different, but equally

number of different possible scenarios, each with its own 20-year trends at various levels of the atmosphere. It is only by performing ensembles of simulations with these models that it is possible to assess whether the observed disparity lies within the range of what should be regarded as physically plausible. Because these numerical experiments are computationally intensive, only a very limited number of them have been run thus far.

It is evident from Figure 2.3 that globally averaged temperature fluctuations associated with El Niño tend to be larger aloft than at the surface, and this behavior is well-simulated in numerical models. These models show evidence of stronger cooling aloft than at the surface in the wake of major volcanic eruptions such as Mt. Pinatubo and in the amplitude of temperature variations induced by fluctuating solar irradiance. The longer the period over which trends are computed, the more these naturally occurring fluctuations in the temperature time series tend to average out. For example, the influence of these phenomena upon the trends should be much smaller when the trends are estimated for a 20-year long record compared with a 5-year record. However, model simulations suggest that such natural variability can still amount to an appreciable fraction of the observed disparity between the global-mean temperature trends at the earth's surface and in the lower to mid-troposphere. Because 20-year trends can be substantially influenced by just a few single or multi-year "warm" or "cold" events, they are not necessarily representative of the true response of the climate system to the more gradual changes in atmospheric composition that are taking place in response to human activities.

A number of different human-induced forcings are, in fact, believed to have contributed to the observed temperature changes during the past 20 years. The climate system is highly non-linear[8] and relatively little is known about the effect on temperature changes resulting from human contributions to the changing three-dimensional distributions of ozone and aerosols, either or both of which may have been partially responsible for the observed discrepancy between surface and lower to mid-tropospheric temperature changes. The aerosol contribution is

---

plausible initial conditions. Differences among the climates in the individual simulations are interpreted as being due to the internal (unforced) variability of the climate system.

[8] Highly non-linear in this context means that there is no guarantee that the response of the climate system to the sum of these forcings would be equal to the sum of its responses to the individual forcings if each of them had occurred in isolation.

particularly difficult to estimate because of the limited understanding of how aerosols affect cloud properties, which affect the transfer of radiation through the atmosphere. In addition to changes in atmospheric composition, land use changes can be a significant factor in causing climate change at the earth's surface.

Despite the many unresolved issues touched on in this chapter and discussed in more detail in chapters 5–9, the progress that has been achieved over the past few years provides a basis for drawing some tentative conclusions concerning the nature of the observed differences between surface and upper air temperature trends, and their implications for the detection and attribution of global climate change.

3

Findings

(1) Surface temperature is rising. Because global warming is a long-term process, which can be masked by year-to-year climate variability, it is most clearly revealed in the longest available record of global temperature—i.e., that of surface temperature (Figure 2.1), which is based on stations determined not to have been substantially impacted by urbanization. In the opinion of the panel, the disparity between satellite and surface temperature trends during 1979–98 in no way invalidates the conclusion of the Intergovernmental Panel on Climate Change report (IPCC, 1996) that global surface temperature has warmed substantially since the beginning of the twentieth century. Accelerated warming during the late 1990s has raised the estimated warming to 0.4-0.8 °C in the past 100 years. The warming of surface temperature that has taken place during the past 20 years is undoubtedly real, and it is at a rate substantially larger than the average warming during the twentieth century.

(2) Based on current estimates, the lower to mid-troposphere has warmed less than the earth's surface during the past 20 years. For the time period from 1979 to 1998, it is estimated that on average, over the globe, surface temperature has increased by 0.25 to 0.4 °C and lower to mid-tropospheric temperature has increased by 0.0 to 0.2 °C.[9]

---

[9] The range of these trend estimates is determined by applying different trend algorithms to the different versions of the surface and tropospheric data sets. Further discussion of the uncertainties inherent in these estimates is provided in chapters 6–9.

(3) Current estimates of surface and lower to mid-tropospheric temperature trends are subject to a level of uncertainty that is almost as large as the apparent disparity between them. The factors contributing to this uncertainty are:

- the uncertainty inherent in temperature trends determined over relatively short time periods;
- the complexity of the algorithms for processing the satellite data, and the limited validation that has been performed on them thus far;
- the possibility of biases remaining in the data sets that have not yet been recognized and corrected;
- the uneven and, in some places, sparse spatial coverage of radiosonde observations and, to a lesser extent, surface observations; and
- the inherent difficulties in correcting for changes in instrumentation and in the siting of radiosonde and surface stations.

(4) The observed trends have been partially, but not fully, reconciled with climate model simulations of human-induced climate change. The simulated three-dimensional spatial pattern of the temperature changes induced by increasing concentrations of a well-mixed greenhouse gas, such as carbon dioxide, is complicated and varies from model to model, but one common aspect is the tendency for the lower to mid-troposphere to warm more rapidly than the surface, except over high latitudes. More realistic model simulations that take into account radiative forcing with combined changes in human-induced and natural factors, including the three-dimensional structure of the changing distribution of ozone, are in better agreement with the observed changes, but they still predict that the lower to mid troposphere should warm at least as rapidly as the earth's surface. The models used to perform these simulations are subject to uncertainties and subject to change as more realistic treatments of physical processes are incorporated into them.

(5) The record of satellite observations of lower to mid-tropospheric temperature is still short and subject to large sampling fluctuations. Recent experiments with a number of different climate models indicate that the inclusion of natural climate forcings such as volcanic eruptions, stratospheric ozone depletion, and solar variability can lead to a broad spectrum of simulated 20-year surface and lower to mid-tropospheric temperature trends. In light of this new information, the panel cautions that trends in such short periods of record with arbitrary start and end points are not necessarily representative of how the atmosphere is changing in response to long-term human-induced changes in atmospheric composition. Given reliable measurements, as outlined in Recommendation #1, the level of confidence that can be attached to the trends will increase as the period of record of upper air measurements lengthens.

(6) It is not currently possible to determine whether or not there exists a fundamental discrepancy between modeled and observed atmospheric temperature changes since the advent of satellite data in 1979. Measurement uncertainties (Finding #3), modeling uncertainties (Finding #4), and sampling uncertainties (Finding #5) were all considered by the panel as possible causes of the disagreement between models and observations. None of them can be singled out as the dominant factor, nor can any one of them be shown to be unimportant. Surface temperature and lower to mid-tropospheric temperature are different entities, which should not be expected to vary in precisely the same manner in response to human-induced and natural climate forcings during a particular 20-year period of record. Hence, the panel concludes that at least part of the observed disparity between the 20-year changes in surface and mid-tropospheric temperature is probably real, but the measurement, modeling, and sampling uncertainties alluded to above make it difficult to precisely attribute the disparity to any particular sources. A more definitive reconciliation of modeled and observed temperature changes awaits the extension and improvement of the observations and the algorithms used in processing them, better specification of the natural and human-induced climate forcings during this period, and improvement of the models used to simulate the atmospheric response to these forcings.

4

Recommendations

In order to monitor global climate change on a decade-to-decade basis in support of national and foreign policy decisions, it will be necessary to better quantify and to substantially reduce the measurement errors inherent in estimates of global-mean temperature, as well as to develop an improved understanding of the processes that contribute to short term variability of global-mean temperature. To achieve these goals, the panel recommends the following actions:

(1) The nations of the world should implement a substantially improved temperature monitoring system[10] that ensures the continuity and quality of critically important data sets. Needed measurements include not only the conventional climatic variables (temperature and precipitation), but also the time-varying, three-dimensional spatial fields of ozone, water vapor, clouds, and aerosols, all of which have the potential to cause surface and lower to mid-tropospheric temperatures to change relative to one another. Management of climate data sets also needs additional attention and support. Raw and processed measurements and follow-on products need to be accessible in a form that enables a number of different research groups to replicate the processing of the more widely disseminated data

[10] The NRC report *Adequacy of Climate Observing Systems* (NRC, 1999) describes characteristics that should be incorporated into the design of climate monitoring systems to facilitate the detection of climate change.

sets and to develop new and improved temperature algorithms. To ensure such access, the ongoing documentation of instrumentation and observing practices, the archiving of data sets, and the provision of raw and processed data sets in electronic form to the scientific community should be regarded as integral parts of the climate monitoring effort and afforded high priority in terms of funding.

(2) The scientific community should perform a more comprehensive analysis of the uncertainties inherent in the surface, radiosonde, and satellite data sets. Such an assessment should involve a detailed analysis of the sensitivity of global-mean temperatures derived from these three different measurement systems to the various choices made in the processing of the raw data—e.g., corrections for instrument changes, adjustments for orbital decay effects in the satellite measurements, and procedures for interpolating station data onto grids. Such studies should also address the comparison of data sets with different sampling characteristics.

(3) Natural as well as human-induced changes should be taken into account in climate model simulations of atmospheric temperature variability on the decade-to-decade time scale. In particular, the studies described in Finding #4 need to be repeated with improved models and with an experimental design that reflects the uncertainties in natural and human-induced forcings.

(4) The scientific community should explore the possibility of exploiting the sophisticated protocols that are now routinely used to ensure the quality control and consistency of the data ingested into operational numerical weather prediction models, to improve the reliability of the data sets used to monitor global climate change.

Part II

Technical Background

5

Introduction

Variations in global-mean temperature are inferred from three different sets of measurements: surface observations, satellite observations, and radiosonde observations. Each of these kinds of measurements has its own particular strengths and weaknesses, as summarized in Table 5.1.

The satellite measurements of tropospheric temperature are the only ones that provide comprehensive global coverage, but rather intricate processing is required in order to infer global-mean temperature trends from the raw radiance data, and these trends have proven difficult to validate independently. Temperature measurements retrieved from the hundreds of balloon-borne radiosonde instruments that are released each day by the various national weather services provide much more detailed information on the vertical structure of atmospheric temperature changes than is available from satellites. The processing of these observations is straightforward, but large gaps in spatial coverage compromise the reliability of global averages, and changes in instrumentation can give rise to spurious trends. Surface temperature measurements derived from thermometers at land stations (housed in instrument shelters) and aboard ships (mostly engine intake temperatures) are more densely spaced than the radiosonde measurements. However, spatial sampling is still an issue in the higher latitudes of the Southern Hemisphere, and ensuring the homogeneity of these data in the face of urbanization and changes in

TABLE 5.1 Summary of the characteristics of surface, MSU, and radiosonde observations.

	Surface	MSU	Radiosonde
Method of observations	Thermometers in shelters (air) or sea water. Since 1982, satellite infrared (IR) oceanic observations blended with in situ observations.	Atmospheric oxygen emits microwave radiation, the intensity of which is measured by the MSU and is proportional to temperature.	Temperature sensors are carried upward through the atmosphere by balloons and the data are radio-transmitted to ground receiving stations.
Spatial coverage of measurements	Good in most inhabited regions and shipping lanes. Sparse elsewhere.	Virtually complete global coverage. Very broad vertical layers (~10 km).	Poor in oceanic regions, in the developing world, and in sparsely populated land areas. Good elsewhere. Good vertical resolution from the surface to the lower stratosphere.
Length of observational record	Beginning in mid-nineteenth century, with expanding coverage in first half of twentieth century. Diminished land station coverage in 1990s.	Begins December 1978.	Beginning in the mid-1940s, with greatly expanded coverage in the early 1960s, but suffering some deterioration in the 1990s
Directness of the temperature measurement	Direct, in situ observation of temperature blended with satellite infrared for sea surface temperature.	Remote measurement of radiative emissions.	Direct, in situ observation of temperature.
Time-varying biases	Raw data are influenced by changes in instruments, observing practices, and land use.	Many biases related to, for example, spacecraft altitude, east-west orbital drift, solar heating, and instrument malfunction.	Many changes in instrumentation, observing methods, and the global network of stations.
Multiplicity of instruments	Sea surface temperature, marine air temperature, and land air temperature measured by tens of thousands of different thermometers of various types.	Usually two spacecraft in orbit; 30,000 observations per day from each; 9 different satellites from 1979 to 1999.	Each sounding made with a new instrument. Dozens of types used over time, varying from country to country, station to station.
Number of independent efforts to produce the data sets	Several groups, employing different methodologies.	One main effort to date.	A few groups, employing different methodologies.

instrumentation and observing protocols has proven to be a major challenge.

To appreciate the issues involved in comparing estimates of surface and lower tropospheric temperature trends, it is necessary to have at least a rudimentary understanding of these three kinds of measurements and the uncertainties inherent in each of them. Chapters 6, 7, and 8 present this basic background information, and the final chapter (9) discusses the issues involved in making comparisons between the different kinds of measurements. Collectively, these last four chapters of the report are the basis for the findings and recommendations presented in chapters 3 and 4.

6

Surface Temperature Observations

SUMMARY OF TRENDS

Globally averaged surface temperatures have been rising over the last century, but at an uneven rate. Temperatures increased from 1900 to the 1940s, and then leveled off or even decreased until the mid- to late-1970s. Since that time, globally averaged surface temperatures have increased again. Sea surface temperatures (SST) and land surface air temperatures both show these same features, although the magnitude of the century-scale changes and year-to-year variability of the land air temperatures are greater than those of SST (see Figure 6.1). The fact that a centennial-scale warming trend with similar decadal-scale features exists in both independently collected data sets serves as a useful check on the reality of the surface temperature trend. Long-term warming, as well as warming over the past two decades, has occurred in both hemispheres and in all seasons (Jones et al., 1999). Although not all regions have warmed, the warming trend since 1976 has been very widespread, as indicated in Figure 6.2. Another widespread feature of the global surface temperature signal is that, at least over the last half century—the period for which we have the most data—the mean daily minimum land air temperature has been increasing at approximately twice the rate of the mean daily maximum temperature (see Figure 9.4) (Karl et al, 1993; Easterling et al., 1997).

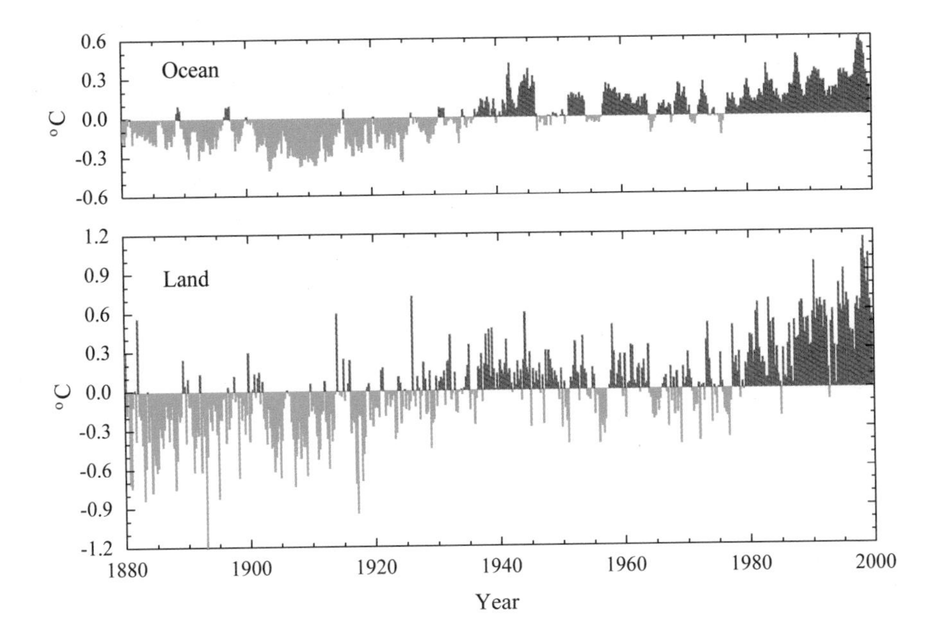

FIGURE 6.1. Time series of seasonally averaged temperature anomalies (1880–1999). The upper figure shows globally averaged sea surface temperature. The lower figure shows globally averaged surface air temperature over land. The anomalies are computed as differences from the 1880–1998 mean. The first season is December 1879–February 1880, and the last season is June–August 1999. The data are based on the Quayle et al. (1999) data set.

The exact magnitude of the temperature trend depends on how the observations are globally averaged. Very different global averaging techniques have been used in various studies. Quayle et al. (1999) created separate globally averaged land and ocean time series using only those grid boxes containing data, and then combined the two series with a 30%/70% weighting, proportional to the global area of land and ocean surfaces. The data used in this approach include satellite-derived sea surface temperatures. Jones et al. (1999) employed a different approach, combining land and ocean in situ data in the same gridded data set, with interpolation into blank grid boxes with at least four neighbors, and then areally averaging the grid boxes into a single time series. In a third approach, Hansen and Lebedeff (1987) produced a global time series

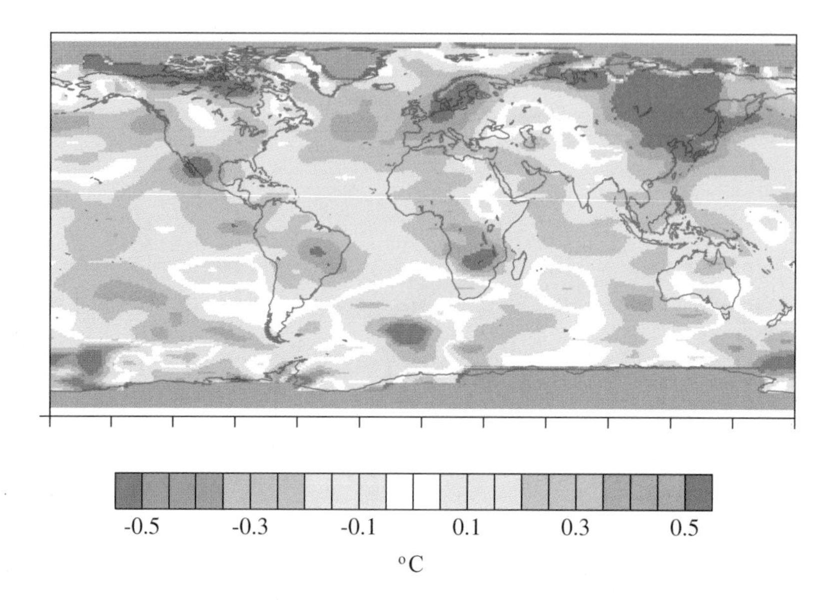

FIGURE 6.2. Global surface temperature trends (°C/decade) for the 20-year period 1979–98 computed using the method of ordinary least squares. The data have been interpolated to a 5°x5° grid by Jones et al. (1999). The gridded data have been subjected to a slight additional smoothing to simplify the pattern.

from only land data by using a given station to represent temperature change to distances suggested by correlation studies. The resulting time series has been shown to be representative of combined land and ocean temperature (Peterson et al., 1998a). This approach has recently been updated to include urban adjustment and an increased number of stations (Hansen et al., 1999). All three of these very different approaches yield quite similar results (see Figure 6.3).

The exact value of the trend in globally averaged temperatures depends not only on which of these methods are used to globally average the data, but also on the time period assessed, as well as on the technique that is used to determine the linear trend. In fact, the method used to determine the linear trend can have a greater impact on the result than the method of creating globally averaged temperatures (Peterson et al., 1998a). The most common method for determining trends—least squared deviations—indicates linear trends from the time series of +0.053,

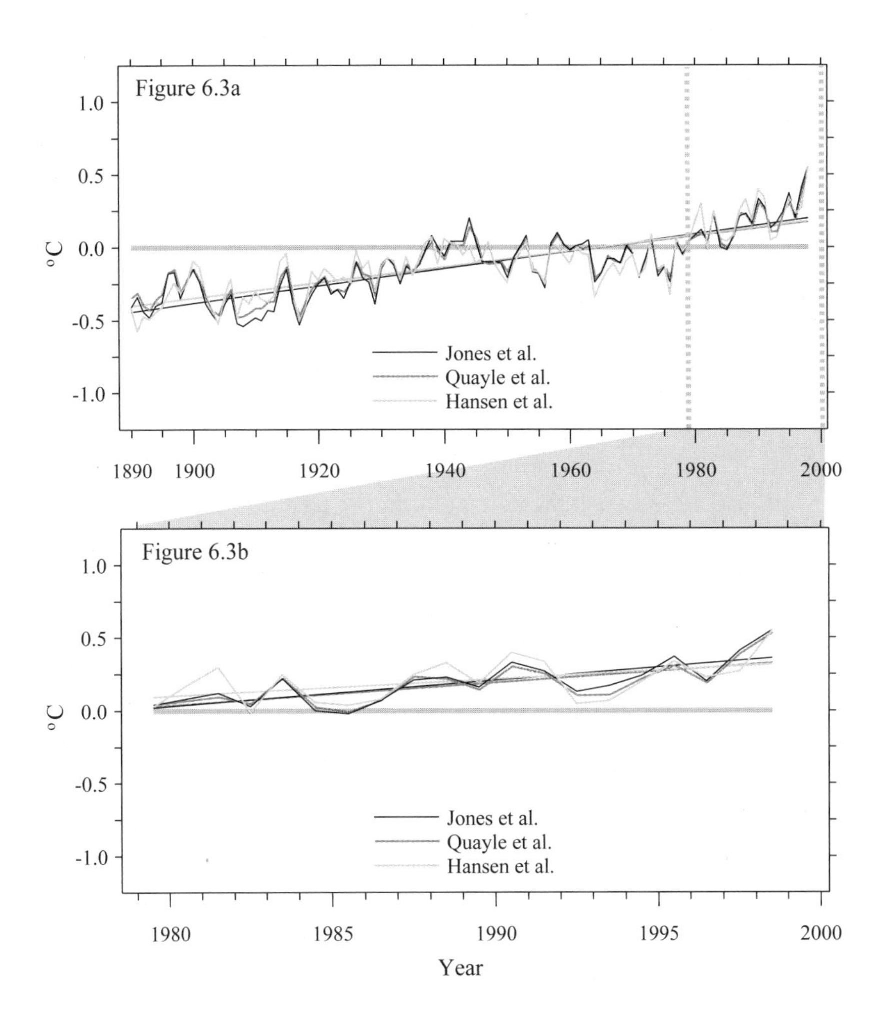

FIGURE 6.3. Comparison of global-mean annually averaged surface temperature time series, illustrating the similarity of results from three different analyses. The series, displayed as anomalies from their 1961–1990 means, are from (a) 1890–1998 and (b) 1979–98, and include least squares linear regression lines for each. The purple line represents the Quayle et al. (1999) data set; the black line represents the Jones et al. (1999) data set; and the aqua line represents the Hansen et al. (1999) data set.

+0.059, and +0.053 °C/decade for the period 1890–1998 and +0.17, +0.19, and +0.13 °C/decade for 1979–98[11] for Quayle et al. (1999), Jones et al. (1999), and Hansen et al. (1999), respectively (Figure 6.3).

As revealed in Figure 6.3, the increase in the global temperature time series is not linear, nor is there any physical basis for expecting it to be linear. Figure 6.3 also shows that 1998 is by far the warmest year on record, with 1997 the next warmest. In addition, the past ten years (1989–98) have been the warmest decade on record.

Considerable corroborating evidence exists to support the analyses indicating that global surface temperature has risen during this period of historical instrumental record. For example, Oerlemans (1994) recorded a general retreat of glaciers around the world over the past 100 years. De la Mare's (1997) analysis of whaling records indicates that a 25% decrease in Antarctic sea ice occurred between the mid-1950s and the early 1970s. There are indications that in the past 20 years, the extent of Antarctic sea ice may have increased slightly, while the extent of Arctic sea ice has decreased (Parkinson et al., 1999; Vinnikov et al., 1999). In addition, measurements from submarines indicate that the average thickness of sea ice across the open Arctic Ocean has declined by 1.3 meters, or 40 percent, from 1958 to the 1990s (Rothrock et al., 1999). Borehole temperatures can also provide an independent instrumental validation of surface measurements. Pollack et al.'s (1998) analysis of underground temperature measurements from four continents indicates that the average surface temperature of the earth has increased by about 0.5 °C in the twentieth century. Warming of surface temperature should also cause a rise in sea level as a result of thermal expansion of the ocean and melting of glaciers. There is considerable evidence that global-mean sea level has in fact risen 10-25 cm over the past 100 years (Warrick et al., 1996).

SOURCES OF UNCERTAINTY IN TREND ESTIMATES

Surface temperature is primarily observed by several thousands of individual thermometers. While the technology for building accurate, reliable thermometers for in situ environmental measurements has been

[11] To further illustrate the sensitivity of a two decade-long time series to the endpoints, if 1976 instead of 1979 is used as the starting date, the trend is approximately 0.02 °C/ decade greater.

in existence for over 150 years, there are many problems with the data from these instruments. The data can be erroneous due to instrument problems (e.g., a bubble in the liquid-in-glass thermometers), or contain errors caused by faulty transcription, digitization, or transmission of the data. Such quality control problems add noise to the data, but are not likely to add a bias to the results because of the large number and variety of sites monitored.

However, homogeneity and spatial coverage problems could potentially add a bias to the results. A surface temperature time series is considered homogeneous if variations are caused only by variations in weather and climate (Conrad and Pollak, 1950). Unfortunately, most long-term climatological time series have been affected by a number of non-climatic factors that make these data unrepresentative of the actual climatic variation occurring over time. These factors include changes in: instruments, observing practices (e.g., depth of the water intake for SST measurements), station locations, formulae used to calculate means, and station environment (e.g., urbanization) (Jones et al., 1985; Karl and Williams, 1987; Gullett et al., 1990; Heino, 1994). They occur in measurements of land air temperature, marine air temperature, and sea surface temperature (Folland and Parker, 1995; Peterson et al., 1998b). Some changes cause sharp data discontinuities, while other changes, particularly change in the environment around the station, can cause gradual biases. All of these inhomogeneities can bias a time series and lead to misinterpretations of the studied climate unless they are accounted for by adjusting or "correcting" the data.

One of the largest sources of uncertainty in global surface temperature analyses is due to incomplete sampling (see Figure 2.7 for data coverage). Large portions of the earth have few in situ observations. This is particularly true of polar regions, uninhabited deserts, and oceanic regions away from the usual shipping and fishing areas. To make matters worse, the locations of observations frequently change with time, and there has not always been a steady improvement in data coverage. For example, starting only in 1978, a limited number of drifting buoys have been placed in the tropical and southern oceans where whaling fleets once took observations. While surface station coverage increased during the nineteenth and most of the twentieth centuries, the difficulties in near-real time international data exchange and the success of many retrospective data gathering efforts (e.g., Bradley et al., 1985; Peterson and Griffiths, 1997) means that we have fewer land surface in situ observations available in the 1990s than in the 1970s or 1980s.

Using the Comprehensive Ocean/Atmosphere Data Set (COADS; Woodruff et al., 1998), Trenberth et al. (1992) analyzed sources of errors for in situ SSTs. By assessing the variability within 2° longitude by 2° latitude boxes within each month for 1979, they found that individual SST measurements are representative of the monthly mean to within a standard error of ±1.0 °C in the tropics and ±1.2 to 1.4 °C outside the tropics. The standard error is larger in the North Pacific than in the North Atlantic and it is much larger in regions of strong SST gradient, such as in the vicinity of the Gulf Stream, because both within-month temporal variability and the within-2° box spatial variability are enhanced. The total standard error of the monthly mean in each box decreases proportionately to the square root of the number of observations available. The overall noise in SSTs ranges from less than 0.1 °C over the North Atlantic to greater than 0.5 °C over the oceans south of about 35 °S.

In addition to the problems of in situ data, satellite-derived SSTs add another source of uncertainty. While Reynolds and Smith (1994) use optimal interpolation to blend satellite-derived and in situ SSTs, there are biases, particularly in areas with sparse in situ data, that can still occur due to volcanic aerosols (Reynolds and Smith, 1994) and differences between satellite-observed ocean skin temperature compared to in situ observed bulk temperature (Reynolds and Marsico, 1993). The incomplete adjustment of satellite data could decrease the linear trend of globally averaged surface temperature between 1979 and 1999 by up to 0.05 °C/decade (Hurrell and Trenberth, 1999). Also, there are large uncertainties in the location of the sea ice margins in regions of sparse data (Hurrell and Trenberth, 1999).

EFFORTS TO CORRECT THE PROBLEMS

Efforts to address these problems started over a century ago. The First International Maritime Conference, held in Brussels in 1853, agreed on the need for international cooperation and adopted a standard set of observational instructions and ship weather-log forms (WMO, 1994). Such efforts have continued through to the present international effort to create the Global Climate Observing System Surface Network (Peterson et al., 1997). At the same time, historical sea surface temperature data continue to be digitized to fill in gaps in the available in situ data (e.g., Woodruff et al., 1998). While new approaches to derive surface temperature over land using satellite data are under development (e.g., Basist et al., 1998), the satellite data these approaches require may only go back to 1987.

The possibility, indeed probability, of erroneous data is addressed by every major data set compiler as part of the quality control effort (e.g., Jones et al., 1999; Peterson et al., 1998c). While all erroneous data points cannot be removed from a data set without the risk of removing a great deal of good data as well, biases due to large isolated errors can be eliminated. Biases due to discontinuities in the observing network are a much more difficult problem to resolve. However, a great deal of work on homogeneity problems has been done over the past decade or more, as summarized in a recent review of homogeneity research (Peterson et al., 1998b). This work attempts to estimate the magnitude of the bias caused by random station moves, installation of new instrumentation, and changes in observing practices such as changing the time of observation of maximum/minimum thermometers from late afternoon to early morning. Once the magnitude of the bias is determined, the data can be adjusted to account for these inhomogeneities.

A more difficult problem is assessing the impact of small, gradual changes in the observing network. Urbanization (and land-use changes in general) and the resultant urban warming is the most commonly cited example of this type of problem. Recent efforts to assess this bias focus on identifying which stations are rural and which are urban, using map-based metadata[12] (Peterson and Vose, 1997) or night-lights derived metadata (Owen et al., 1998). Long-term global temperature trends calculated both from the full land surface network, and from rural stations only, turn out to be very similar (differing by about 0.05 °C per 100 years), despite some differences in regions sampled (e.g., India has few long-term rural stations) (Peterson et al., 1999).

The uneven spatial distribution of in situ data, and the change in their distribution over time, can also potentially create biases. Some of the approaches to addressing this problem are: (a) acquiring more data through digitization of historical records, (b) improving international

---

[12] Metadata (or data about data), in this context, is information that describes the environment in which a measurement is made and/or the methods and/or tools used to make the measurement.

exchanges, (c) reconstructing full global grids using the spatial and temporal covariance of the field (e.g., Smith et al., 1998), and (d) developing new space-based observing systems. However, global coverage of in situ data can never be achieved, particularly historically. Therefore, inventive area averaging techniques have been developed to provide robust estimates of global temperatures. These techniques include grid-box averaging of climate anomalies (e.g., Jones, 1994), or averaging of the interannual change in temperature (Peterson et al., 1998a). A more complex approach that interpolates anomalies adjusted to regional reference stations produces information for each grid box (Hansen and Lebedeff, 1987). Smith et al. (1998) also fill in the full grid using the spatial and temporal covariance of the sea surface temperature field together with the available data. Within this latter approach is the assumption that the covariance pattern developed in the satellite era is an appropriate guide for interpolating data in earlier eras.

Several efforts have been made to put error bars on global surface temperature time series, primarily by focusing on the impact of inadequate spatial sampling and using model simulations of global climate. Jones et al. (1997) estimated that the typical standard errors for annual data on the interannual time scale since 1951 are about ±0.06 °C.[13] Errors associated with century-scale surface temperature trends are probably an order of magnitude smaller than the observed warming of about 0.5 °C per 100 years since the late nineteenth century (Karl et al., 1994).

[13] Unless stated otherwise, the quantitative error estimates given in this report represent 95% confidence intervals.

7

MSU Observations

INTRODUCTION

The Microwave Sounding Unit (MSU) is a microwave radiometer that flies aboard NOAA's polar orbiting weather satellites. Each day, the MSU observes approximately 80% of the earth's surface, with the orbit shifting slightly each day so that 100% coverage is achieved over a three to four day period. To date, nine MSUs have been used operationally, forming an uninterrupted daily time series from 1979 to the present. One MSU remains in orbit as the last representative of this series, which in 1998 began to be replaced by the Advanced Microwave Sounding Unit.

The MSU observes the earth's natural upwelling radiation at four frequencies between 50 and 60 GHz. The particular channel upon which this report focuses is channel 2 (53.74 GHz). The radiation detected by channel 2 comes from the earth's atmosphere (90-95%) and surface (5-10%). The bulk of this radiation originates in the troposphere, the layer from the earth's surface up to about 10 km. The MSU also monitors the lower stratosphere through its channel 4. The intensity of radiation observed in these channels is directly proportional to the temperature of the air; hence, MSU can be used as a satellite "thermometer" for measuring air temperature.

The air temperature computed directly from channel 2 is representative of the middle-to-upper troposphere (centered about 7 km above the surface). A small but significant part of this radiation emanates from the lower stratosphere. This is problematical for detecting changes related to greenhouse warming, because the warming in the troposphere is expected to be accompanied by cooling in the lower stratosphere. Therefore, the blending of the radiation from these two layers seen by channel 2 partially or completely damps out any greenhouse warming signal.

There are two approaches to processing the observations to obtain a pure tropospheric temperature. One approach is to combine observations from the different channels of the MSU. The other approach is to exploit observations from the different scan angles using MSU channel 2 alone. The latter approach is the one discussed in this report because it is much more mature in terms of the extent of data-set development and validation. To obtain a temperature closer to the earth's surface in the latter approach, observations from different scan angles that the MSU uses to view the earth's atmosphere are arithmetically combined to reduce the influence of the upper troposphere and stratosphere (Spencer and Christy, 1992). The advantage of this technique is that the resulting lower tropospheric temperature (often referred to as "MSU 2LT") is closer to the earth's surface (centered about 4 km high). Because the central issue being examined in this study is the expectation by some that the lower troposphere should exhibit similar temperature trends as the surface, the panel exclusively discusses the MSU 2LT product in this report. The drawback of using the MSU 2LT product is that the retrieval method relies on a subtraction of adjacent view angles, which (a) increases measurement noise and (b) doubles the sensitivity of the measurements to surface emissions (to 10% over oceans, 20% over land) (Spencer and Christy, 1992). These effects more than double the error characteristics for MSU 2LT relative to MSU 2.

MSU TEMPERATURE TRENDS

The geographical distribution of MSU lower to mid-tropospheric temperature trends is shown in Figure 7.1. It is evident that the regions of rapid surface warming apparent in Figure 6.2 (e.g., Western Europe, Eastern Russia) tend to be characterized by rapid warming aloft, and vice versa. In contrast to the surface data, which exhibit a warming trend at most locations, the satellite data show roughly equal areas of warming and cooling. Whereas it is obvious from a visual inspection of Figure 6.2

that the global-mean surface temperature trend is upward, the same cannot be said of the trend inferred from the MSU data; the global-mean trend may be viewed as a small difference between warming trends over some areas and cooling trends over others.

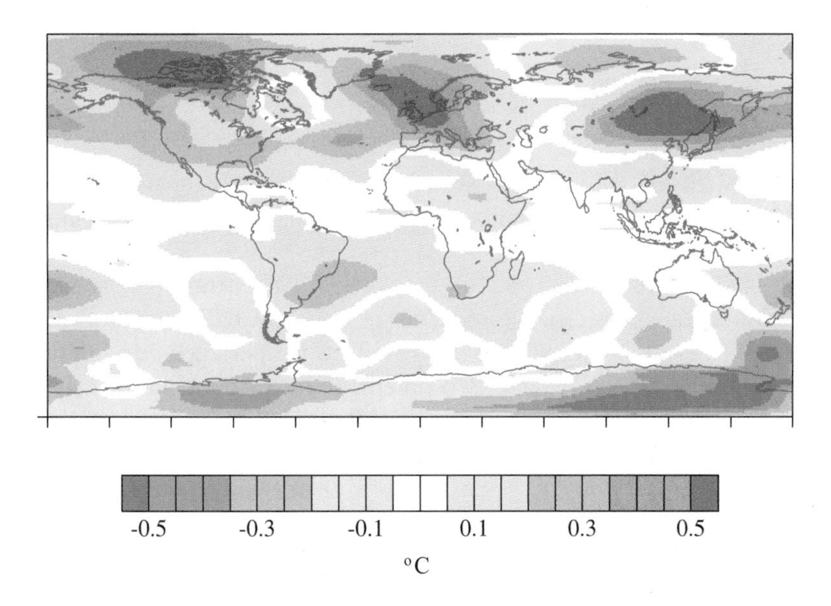

FIGURE 7.1 Global lower to mid-tropospheric temperature trends (°C/decade) from the MSU version D over the 20-year period 1979-98. These ordinary least square trends are computed from data from Christy et al. (2000).

The time series of seasonally averaged global MSU lower to mid-tropospheric anomalies is shown in Figure 2.3. The data reveal major swings in temperature over relatively short periods, which are largely the result of climate perturbations such as the El Niño–Southern Oscillation (ENSO) warming events in 1983, 1987, 1991, and 1997; the cooling events in 1985, 1988, 1996, and 1999; and the volcanic aerosol cooling events in 1982 and 1991 (Christy and McNider, 1994). Over the past 20 years, MSU observations indicate that the globally averaged lower to mid-tropospheric temperature has increased at a rate of approximately 0.05 °C/decade, as computed using the ordinary least squares statistical method. It is important to note that a single 20-year period of record is unlikely to be representative of both future rates of change and rates of change over much longer periods of the historical record. The reasons for this include the large natural variability associated with El Niño, together with both long- and short-term changes in the external factors that can influence climate, such as volcanic eruptions, the sun, and greenhouse gases.

SOURCES OF UNCERTAINTY IN TREND ESTIMATES

As with all climate data sets, there are drawbacks and limitations associated with the MSU temperatures. To begin with, MSU channels 2 and 4, as well as products derived directly from these channels, have quite coarse vertical resolution. The MSU cannot measure the temperature at a specific altitude, as is done by balloon observations. Rather, it detects a weighted average of the temperature throughout the atmosphere. This is particularly problematic for diagnosing the causes of climate change, because the different atmospheric layers may exhibit different long-term temperature trends. For example, during the 20-year period of MSU operation, the evidence is clear that the surface layer has experienced a warming trend, while significant cooling has occurred in the stratospheric layers observed by MSU channel 4.

Systematic measurement errors are another problem for the MSU. Variations in sensor gain (i.e., the ratio of the perceived signal to the actual signal) are particularly problematic in that these gain variations can be misinterpreted as trends in air temperature. In principle, gain variations can be measured (and hence corrected) by the onboard two-point calibration system, consisting of a warm load at a known temperature and cold space observations at 2.7 Kelvin. However, in practice the gain cannot be exactly determined because of the presence of small non-linearities in the MSU response to incoming radiation. It appears that this non-linear response was not properly characterized in the pre-launch thermal-vacuum tests. As a result, the two-point calibration system is not completely removing gain variations. To correct this problem, a rather elaborate post-launch analysis relies on the fact that the radiometer gain varies with the physical temperature of the front-end radiometer components (Christy et al., 2000). Systematic measurement differences between different satellites over many years are correlated with the physical temperature of the radiometers. A time-varying gain correction, which is a function of instrument temperature, is then applied to those MSUs that appear to have experienced calibration problems. This additional complexity in MSU data processing, in conjunction with a non-linear gain problem that is poorly understood, decreases our confidence in the ability of MSU to measure long-term trends. To the extent that these problems are random, they can be reduced by averaging the millions of observations (over 15,000 per day). However, residual systematic calibration errors not correlated with the radiometer temperatures cannot be removed in this way.

Systematic measurement errors also impact our ability to intercalibrate the series of MSUs. The offsets between the MSU observations from each of the satellites can be readily determined if the bias is constant, but is more difficult to determine if the satellite is drifting in orbit (see below). During periods of satellite overlap, the temperatures measured by two different MSUs are compared. Typically, a temperature offset of up to ±0.4 °C is found. Given the fact that the MSUs are nearly identical instruments, it is not entirely clear what is causing these offsets (perhaps very small manufacturing differences or satellite altitude differences). In any case, these inter-satellite offsets are addressed by adding small bias corrections to data from the various MSUs so that they are, on average, in agreement during overlap periods. This type of satellite intercalibration works best for long overlap periods (one year or greater). Unfortunately, in the case of the NOAA-9 MSU the overlap period was only 102 days. To bridge the NOAA-9 period, several different adjustment methods were tested (Christy et al., 1998). These methods produced a spread in trends of about 0.1 °C/decade. The method that produced the lowest error characteristics and the most data available for analysis was chosen. This selected method also had the desirable feature of producing a decadal trend that was close to the mean of all other possible methods. Nevertheless, the relatively small number of observations during overlapping periods in 1986–87, coupled with uncertainties arising from the choice of method used to correct inter-satellite biases, introduces a further source of uncertainty in MSU-based estimates of decadal-scale trends.

Aside from the issue related to the method chosen for merging the satellite data, there is also an uncertainty associated with determining each satellite's bias relative to some reference value. Several tests have been performed in which the biases were calculated from separate subsections of the overlapping periods, demonstrating very high reproducibility of the results. Even so, small inter-satellite bias errors can accumulate in such a way as to introduce errors in the long-term trend.

Even in the absence of measurement error, the drift of the satellite orbit has the potential to introduce spurious signals into the MSU temperature trend. One component of orbit drift is the decrease in satellite altitude that occurs after launch. Fortunately, this effect can be precisely modeled using the satellite orbital data and a relatively simple radiative transfer model. However, it is worth noting that this particular effect, known as orbit decay, was not recognized until quite recently (Wentz and Schabel, 1998), which suggests that there may be other subtle but important corrections that still need to be applied. The other component of orbit drift is the change in the local time for satellite observations. As the satellite slowly drifts in time, it will observe a warming or cooling trend simply due to the change in time of day being observed on earth (night is cooler than day). If no correction is applied, then this diurnal signal will be confused with an interannual signal because the diurnal drift is on time scales of the order of 0.5 hr/year. As was the case for the radiometer gain problem, a complex analysis is required to remove the diurnal drift signal. In Christy et al. (2000), the effect of the diurnal drift is estimated from the difference between the left and right sides of the MSU viewing swath, which represents a difference in local time ranging from over one hour in the tropics to several hours at the poles. This procedure attempts to remove most of the diurnal signal, but some error will remain. The preceding discussion of problems addresses each individually, whereas in practice these problems are not necessarily independent (for example, a bias that is changing with time), increasing the uncertainty of the corrections.

In light of the aforementioned problems, the obvious question is how accurately can MSU measure long-term trends. This is a difficult question to answer. The errors associated with radiometer gain, inter-satellite calibration, and diurnal drift are difficult to model, and there is always the possibility of other, yet to be found, effects.

The most recent version of MSU 2LT, which includes adjustments for orbital changes, instrument heating, and changes in diurnal sampling, is referred to as version D, distinguishing it from earlier versions labeled A, B, and C. Over the entire time series, the adjustments to version D relative to version C affect the trend of version C as follows: (1) orbit decay, +0.10 °C/decade; (2) diurnal drift, -0.03 °C/decade; and (3) instrument body effect on several instruments, along with the impact of new NOAA-12 calibration coefficients, -0.04 °C/decade. The net effect therefore is +0.03 °C/decade of version D versus version C.

The time series of globally averaged temperatures from versions C and D are compared in Figure 7.2 together with the difference time series (D minus C), which represents the correction to version C. During individual seasons, the corrections amount to as much as a few tenths of a degree C. The spikiness in the difference between C and D in post 1991 data is due to the erroneous calibration coefficients used in version C, discovered by Mo (1995), and corrected in version D. The upward trend in the corrections, which amounts to +0.03 °C/decade, is evident from a visual inspection of the difference time series in Figure 7.2. The cumulative effect of all the corrections that have been made to the MSU data since the release of version A also amounts to +0.03 °C/decade.

Consistent with the recommendations of this report that independent processing efforts should be undertaken, a separate MSU time series has been created by Prabhakara et al. (1998). This data set is based on a small portion of the MSU channel 2 data, with no adjustments for the effects described above. In many ways it is similar to the Spencer/Christy MSU 2 version A. However, the data set has yet to be completed for 1979-98, and has not been compared with other data sets in detail for assessment purposes.

Because there is only one set of observations and a limited number of processing efforts, there is no rigorous way to objectively compute the MSU's measurement precision. However, assessments can be made of the effect of perturbing the methodology, as well as the assumptions that are used to compute MSU temperatures. One analysis of this type, which also included direct radiosonde comparisons, suggests a measurement error of ±0.06 °C/decade[14] for the MSU trend (Christy et al, 2000).

---

[14] This particular value was derived from three separate calculations. (1) The 95% error estimate for each of the three correction procedures was determined and applied to the data. Then the year of worst reproducibility was identified (i.e., to create a conservative estimate). The magnitude of this error was applied to all years (i.e., each year had an error bar with which the 95% trend range could be determined.). (2) Using co-located radiosonde and MSU differences on 2.5-degree grids, error estimates were calculated for regions and then scaled globally in a conservative manner by assigning all error to MSU (because these were stable, U.S. controlled stations). (3) Using two different global radiosonde data sets, error estimates were derived. The value ±0.06 °C/decade encompasses the 95% range from these three methods of estimation. However, this estimate does not include testing either of the sensitivity of the period analyzed, or of the substantial uncertainty associated with adjustments to the data from NOAA-9.

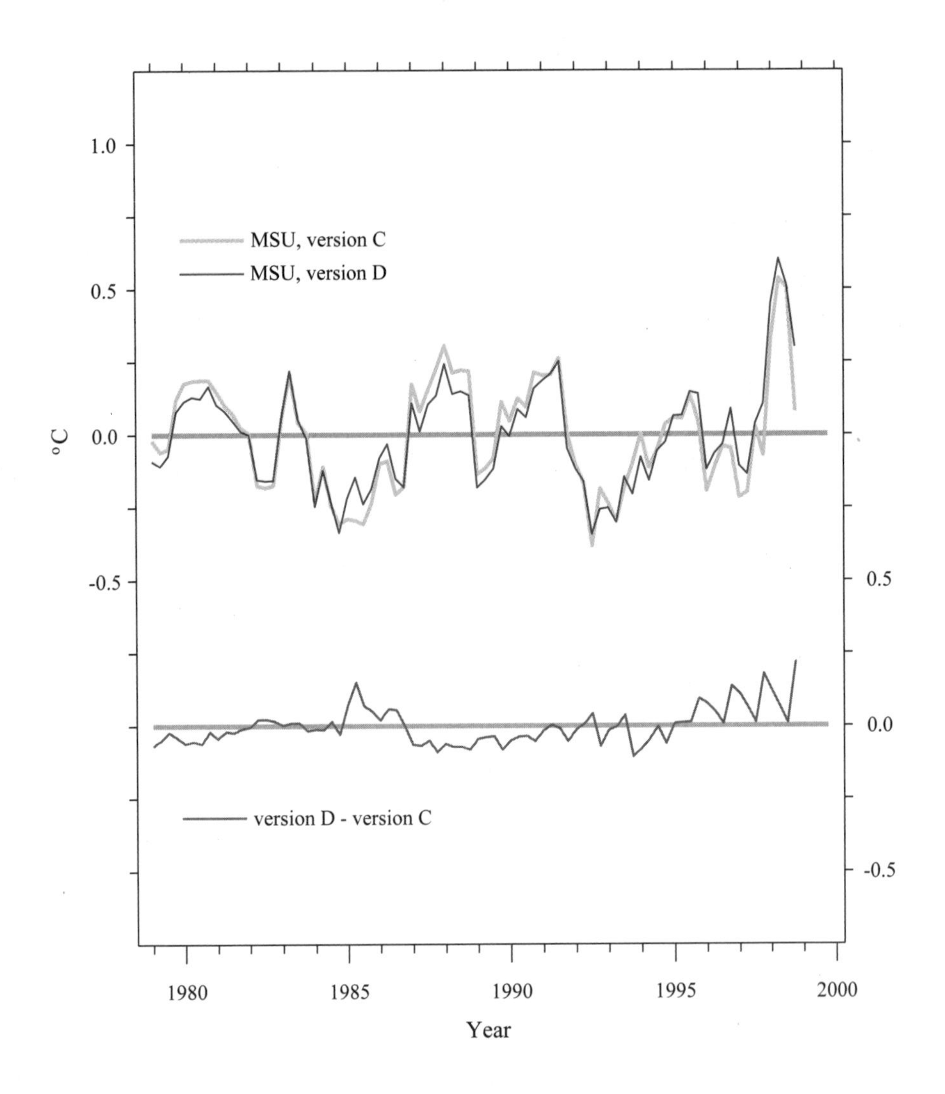

FIGURE 7.2 Globally averaged time series of lower to mid-tropospheric MSU temperature anomalies from version C (orange curve; Christy et al., 1998) and version D (red curve; Christy et al., 2000), as well as the difference between versions D and C (D-C) (gray curve, bottom) from 1979 to 1998. To highlight the differences between the curves, the vertical scale has been expanded by 50% relative to the report's other MSU time series figures.

Others, however, view this analysis as not rigorous enough to reliably identify measurement error at the precision required for decadal-scale climate monitoring, and estimate the measurement error in the MSU trend to be about ±0.1 °C/decade (Hurrell and Trenberth, 1998), or possibly greater.

8

Radiosonde Observations

SUMMARY OF TRENDS

Several data sets compiled from radiosonde observations have been used to monitor atmospheric temperature trends (Angell and Korshover, 1975; Angell, 1988; Parker et al., 1997). Recent analyses of various versions of these data sets indicate slight warming trends of up to 0.1 °C/decade or more in lower tropospheric temperature for the period during which MSUs have been operational (1979–98). Exact trend values vary depending on the data source, treatment, and trend-fitting method (e.g., Angell, 1999; Parker et al., 1997; Santer et al., 1999; Santer et al., 2000).

SOURCES OF UNCERTAINTY IN TREND ESTIMATES

There are several unresolved challenges in determining reliable, global, radiosonde-based temperature trend estimates. These challenges are outlined below, together with some of the methodologies that have been used to address them.

Background

Radiosondes have for several decades been the primary means of obtaining atmospheric vertical profile data from the surface to the lower stratosphere. They are routinely used as input to the operational meteorological analyses that are used in numerical weather prediction and meteorological diagnostics. In the absence of other in situ measurements, radiosonde observations have recently been used to assess trends of atmospheric conditions above the surface, even though they were not designed for this purpose. The instrument packages carried aloft by balloons are generally equipped with temperature, humidity, and pressure sensors, whose measurements are radio-transmitted to a ground receiving station. Wind data are also obtained by tracking the position of the instrument during ascent. Temperature sensors vary according to the manufacturer and model of the radiosonde; most contemporary instruments carry a thermocapacitor, wire resistor, thermocouple, or bimetallic sensor. An excellent overview of radiosonde instruments, including discussion of measurement error characteristics, is provided by the World Meteorological Organization (WMO, 1996).

Currently, the global radiosonde network nominally includes about 900 upper-air stations, of which about two-thirds make observations twice daily (at 0000 and 1200 Coordinated Universal Time (UTC)). The network is predominantly land-based and favors the Northern Hemisphere (Figure 2.6). Radiosondes can achieve heights of about 35 km, although many soundings terminate below 20 km because less expensive balloons burst at a lower altitude. There has been some deterioration in the radiosonde network in recent years. The loss of navigational systems used to track the sondes has led to at least temporary closing of some stations, particularly in Africa. Efforts to reduce operating costs have led to station closures and reduced observing schedules in some parts of the former Soviet Union and elsewhere.

Measurements are made and transmitted with approximately 10-50 m resolution during ascent, but archived sounding data may contain only about 20 data levels per sounding. Radiosonde-based data sets for climate monitoring come from two basic data products: individual soundings containing all reported data (Angell, 1988), or monthly mean data (known as CLIMAT TEMP reports) at mandatory pressure levels only (Parker et al., 1997). Only about 45% of stations provide CLIMAT TEMP reports in addition to the daily sounding data. Missing daily data can cause substantial random errors in estimates of monthly means, especially if the available data are unevenly distributed in time.

Data Homogeneity Problems

Sampling Changes: The spatial and temporal characteristics of radiosonde observations have tended to change through time, in part because these observations are generally made for operational weather analysis purposes rather than long-term trend detection. Spatial biases may be introduced through changes in the number, location, and characteristics of radiosonde sites. Further exacerbating this problem is the fact that the land surface characteristics of radiosonde sites may change through time, biasing surface and boundary layer observations. Shifts in site locations of even a few kilometers can have a similar effect. In addition to spatial sampling issues, biases can also be introduced if the diurnal sampling time or frequency changes. For example, observing times were not fixed at 0000 and 1200 UTC until 1957, making data from earlier years potentially biased relative to more recent observations. Even today, measurements are not always conducted twice daily at all stations.

Instrument Changes: There have been many and widespread changes of radiosonde sensors during the history of the global radiosonde network. These changes often brought useful improvements in precision and accuracy, essential for weather analysis and forecasting, but they also prejudiced the homogeneity of the records from the perspective of climate change analysis (Gaffen, 1994). For example, efforts to update the sonde's temperature-sensing technology and efforts to mitigate the effects of solar radiation on the sondes have, in some cases, introduced time varying biases. Parker and Cox (1995) documented an increased dominance of radiosondes made by the Vaisala company in recent years. Since their paper was published, many North American stations have also switched to Vaisala instruments, although stations in Russia, China, and Japan continue to use national instrumentation. Within each class of radiosondes (such as Vaisala), there have also been progressive developments (e.g., the Vaisala RS11 through RS90 series), which have introduced heterogeneity. Another problem is that the documentation of instrument types and the timing of instrument changes is not always complete or accurate (Gaffen, 1993).

Data Treatment Changes: Changes in the way raw radiosonde observations are processed can also have a significant impact on the long-term record. Changes in corrections applied to the temperature data (to reduce errors resulting from solar and infrared radiation impinging on the sensor and errors resulting from the time lags in instrument response as the sensor ascends through the atmosphere) are detectable in the data (Gaffen, 1994). However, these effects are more noticeable in the data from the upper troposphere and lower stratosphere than from the lower troposphere, and more noticeable early in the radiosonde record than in recent years (the time of overlap with MSU observations). A third type of change in data treatment affects the CLIMAT TEMP monthly averages, but not the individual sounding data. Changes in the rules by which stations compute their monthly averages, including which observing time to use and how many days of data must be available, can have large effects, which are revealed by comparisons with monthly averages computed using a consistent set of rules (Gaffen et al., 2000).

Variety of Methods of Estimating Global Trends in Layer-Mean Temperatures

Methods for Obtaining Layer-Mean Temperatures: The MSU temperature product discussed in the previous chapter is a vertically-broad and non-uniform representation of tropospheric temperature. Therefore, comparisons with radiosonde data are most meaningful if the radiosonde data are processed to represent the same portion of the atmosphere as the MSU product. At least two different techniques have been used. The simplest involves computing the mean mid-tropospheric temperature, weighting all levels (e.g., 850 to 300 hPa) equally. The disadvantage of this method is that it does not reflect the unequal contributions from each of the levels that underlie the MSU product. In the second technique, radiosonde temperature data at different altitudes are weighted to more closely resemble what the satellite would have observed. However, this method, termed the 'static weighting method,' does not account for variations in atmospheric moisture as a function of space or time. The biases associated with this technique are, however, not large relative to other sources of bias in the global time series (Santer et al., 1999).

Methods for Calculating Global Average Temperature Anomalies: Estimates of global and regional temperature anomalies depend on the selection of stations, the method of averaging station anomalies, the method of gridding, and the method of averaging gridded values, which are discussed below.

For the reasons mentioned previously in the *Sampling Changes* section, fixed networks smaller than the full observing system have been chosen for the determination of temperature trends. Angell and Korshover (1975) selected 63 stations in their pioneering efforts to develop a global temperature monitoring capability. The Global Climate Observing System / Baseline Upper-Air Network designates approximately 150 stations for the same purpose. Other efforts (e.g., Oort and Liu, 1993, Parker et al., 1997) attempt to incorporate data from as many stations as possible with an aim of maximizing spatial and temporal coverage. That approach, however, suffers from the inconsistencies that are introduced into the network through time.

An annual temperature anomaly of the selected stations may be calculated as the average of the available months, or as an average of the available seasons. If the record is incomplete in a systematic manner, the weighting implicitly applied to individual monthly data may introduce biases. Similar considerations apply to the calculation of monthly statistics from daily data, and these are especially relevant when a month's data consist of a different number of daytime and night-time ascents.

To grid the station data, anomalies within a gridbox may either be weighted equally, or weighted according to the distance of the station from the center of the gridbox (e.g., Parker et al., 1997). Some schemes also fill unsampled gridboxes using eigenvector-based reconstructions (Parker et al., 1997), objective analysis schemes such as "Conditional Relaxation" (Oort and Liu, 1993), or optimum interpolation (used on sea surface temperatures by Reynolds and Smith, 1994). The validity of interpolations decreases away from the observation sites in proportion to the spatial decorrelation scale of the temperature data. Finally, model-based reanalyses, which also incorporate surface and satellite observations, use complex data-assimilation schemes based upon known physical relationships (e.g., geostrophy) and detailed statistical quality controls (Kalnay et al., 1996; Uppala, 1997).

There are many ways of averaging gridded temperature anomaly values into global indicators. Although each of these has its strengths, the problem of incomplete spatial coverage remains nonetheless. The most direct method is to average all of the gridded anomalies, weighting each according to the area of the gridbox (Parker et al., 1997). This approach naturally places more weight on the Northern Hemisphere where there are more observations. Alternatively, the grid boxes may first be averaged into larger grid boxes before averaging globally (e.g., Parker et al., 1997). In data-sparse regions, this procedure gives greater weight to isolated boxes where data happen to exist, and was shown by Santer et al. (1999) to have a noticeable impact on trend estimates. Another method is to calculate latitudinal averages and then to average these bands (e.g., Angell, 1988). For the tropics, where data are sparse but also where temperature variations tend to be spatially coherent, this form of averaging offers some potential advantages (Wallis, 1998). On the other hand, it can introduce large errors if relatively few stations are operating within a given latitudinal band, and if their temperatures are not representative of the latitudinal average.

EFFORTS TO CORRECT THE PROBLEMS

It is only in the past few years that serious attempts have been made to adjust radiosonde data to remove the effects of artificial changes, and several different approaches have been proposed. Because of the demonstrated sensitivity of trends to data adjustments, and the distinct possibility that some adjustments may introduce more error than they remove, it will be important to compare adjusted data sets and their effects on trends in the future.

Parker et al. (1997) used MSU retrievals as references to test for heterogeneities in temperatures at individual radiosonde stations since 1979 (the beginning of the MSU record), and to make adjustments if necessary. MSU channel 4 and the 2LT retrieval were used for the stratosphere and troposphere, respectively. Adjustments were only made in cases of known instrumental or procedural changes at the radiosonde stations. Some changes in instrumentation had resulted in spurious cooling of up to 3 °C in the lower stratosphere, but biases were much smaller in the troposphere.

Radiosonde temperatures can be adjusted independently of MSU data using semi-empirical models of the thermodynamics of radiosondes, and the results verified using day-night (strictly, 0000-1200 UTC) differences (Luers and Eskridge, 1998). These models take into account known changes in the rate of ascent and in observing time, as well as changes in sensors. These models are now being applied to the majority of radiosonde types used since 1960.

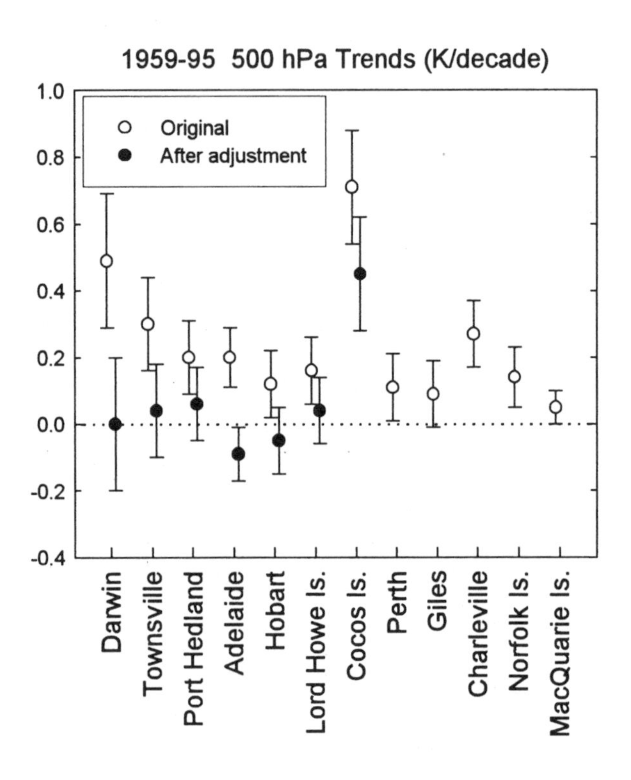

FIGURE 8.1. Mid-tropospheric (500 hPA) temperature trends (and their 95% confidence intervals) at twelve radiosonde stations operated by the Australian Bureau of Meteorology for 1959-1995 (Gaffen et al., 2000; reprinted with permission of the American Meteorological Society). The temperature data for seven stations were adjusted to account for a 1979 change in radiosonde instrument type from the Astor Mark I sonde to the Phillips Mark II sonde.

A third approach utilizes statistical methods to objectively identify abrupt shifts, or change-points, in time series. Gaffen et al. (2000) experimented with two different statistical approaches, representing opposite extremes. One, which relies only on statistical identification and adjustment of the data, is very liberal in that it cannot distinguish between artificial and natural variability. The other, which incorporates station history metadata, is very conservative in that it adjusts only for artificial changes which are identified with a high degree of confidence. These experiments demonstrate that adjustments for change-points can yield very different time series and trends, depending on the scheme used to make adjustment and the manner in which it is implemented. This is illustrated in Figure 8.1, which shows mid-tropospheric (500 hPa) temperature trends from twelve stations operated by the Australian Bureau of Meteorology (Gaffen et al., 2000). The trends in the original data for the period 1959–95 show warming of between 0.05 and 0.71 °C/decade. The data from seven stations were adjusted due to a step-like warming of approximately 0.75 °C associated with a 1979 change in radiosonde type. The effect of the adjustment is to substantially reduce the trends and in some cases to change the warming to a cooling.

Model-based reanalyses (see the previous discussion on gridding radiosonde data) offer a further potential means of radiosonde temperature bias detection and removal through comparisons with first-guess fields.

Each of these strategies for radiosonde data adjustment, except the last one, depends to some degree on metadata—information about the history of instruments and observing practices at each station. Despite recent efforts to compile and digitize global radiosonde metadata (Gaffen, 1993, 1996), there are gaps and uncertainties in the historical information. Current efforts to collect and maintain metadata archives are minimal and should be enhanced.

9

Trend Comparisons

COMPARISONS BETWEEN MSU AND RADIOSONDE DATA SETS

One means of assessing the accuracy of the MSU data set is to compare its results with those of other independent measurement systems. Strictly speaking, it is impossible to compare time series of global-mean tropospheric temperature anomalies based on satellite data with radiosonde measurements, because the radiosonde network does not provide global coverage (see the discussion in the Radiosonde Observations chapter). To illustrate this point, Figure 9.1 compares two global-mean tropospheric time series based on the same MSU measurements. The red curve represents full global coverage, while the pink curve is based on just a limited sampling of grid points designed to mimic the existing distribution of radiosonde stations. The least squares trend of the global coverage time series is 0.06 ±0.11 °C/decade, compared with 0.14 ±0.10 °C/decade for the sub-sampled time series. The differences between these two curves indicate that the radiosonde network may not be sufficiently dense to provide reliable estimates of global-mean temperature anomalies.

Although one cannot compare MSU and radiosonde-based time series of complete global-mean tropospheric temperature anomalies, it is nonetheless possible to compare the sub-sampled MSU time series in Figure 9.1 with the corresponding radiosonde-based time series. In

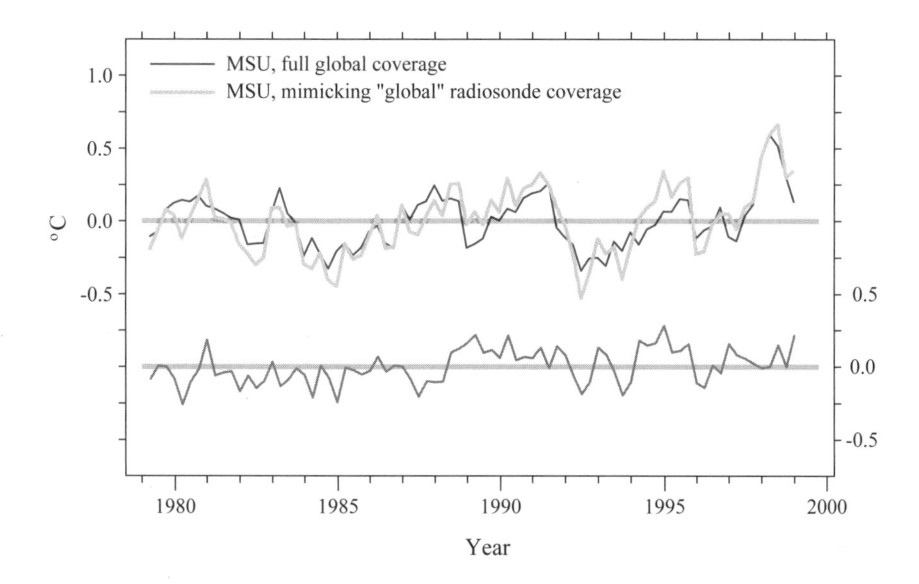

FIGURE 9.1 Global-mean seasonally averaged tropospheric time series based on the same MSU 2LT measurements (Christy et al., 2000). The red curve represents full global coverage, while the pink curve is based on a limited sampling of grid points designed to mimic the existing distribution of radiosonde stations. The dark gray curve (bottom) represents the difference between the sampled and full data sets. The light gray lines represent the means of each time series. The first season is March–May 1979 and the last season is December 1998–February1999. The sub-sampled MSU data were supplied by the U.K. Meteorological Office (UKMO).

principle, a close match constitutes an independent verification of the MSU data. The two curves, shown in Figure 9.2, exhibit a number of common features and rather similar trends.

Uncertainties exist in assigning confidence levels to trends because of persistence in the data, which may or may not be due to the trend itself. There is no unique set of confidence intervals for the relatively short atmospheric temperature time series considered here. The estimated confidence intervals depend on the underlying statistical model that is used to describe the data, as well as on the exact period considered and the sampling interval (i.e., whether one uses monthly, seasonal, or annual means). One approach, following Cryer (1986), yields trends of 0.14 ±0.10 °C/decade for the subsampled MSU data and 0.04 ±0.07

°C/decade for the radiosonde data (see Hurrell et al., in review). Other approaches suggest even larger confidence intervals (Santer et al., 2000).

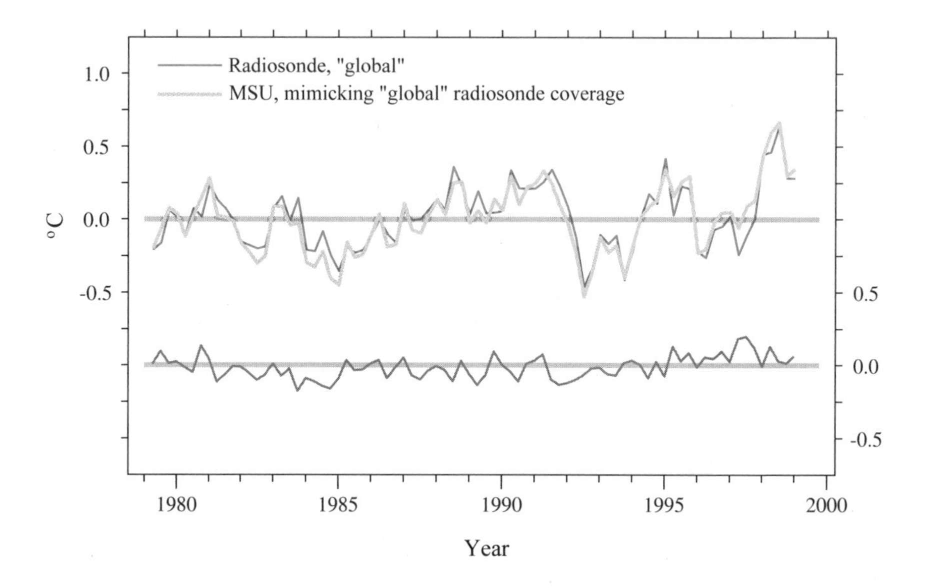

FIGURE 9.2 Equivalent 'global-mean' MSU 2LT and radiosonde tropospheric time series. The pink curve in this figure is identical to the pink curve in Figure 9.1 and represents MSU data from those grid points from which long-term radiosonde observations are available. The green curve in this figure is the average of the radiosonde data (Parker et al., 1997) at the same grid points. The dark gray curve (bottom) represents the difference between the MSU and radiosonde data sets shown in this figure. The axes are the same as in Figure 9.1. Both data sets were supplied by the UKMO.

Some of the discrepancies between the MSU- and radiosonde-based time series are a consequence of changes in radiosonde instrumentation that have not been corrected. Christy et al. (2000) performed a comparison analogous to the one shown in Figure 9.2, but for a subset of 97 radiosonde stations operated by the United States, whose records are believed to be free of such artificial discontinuities for a specific period of time (Figure 9.3). The level of agreement between the radiosonde data and the MSU data was moderately improved by restricting the comparison to this more limited selection of stations. High, middle, and low latitude subsets of these stations (not shown) exhibited a comparably high level of agreement with the MSU data. For the 97 stations as a whole over the 16-year period from 1979 to 1994, the station-by-station root mean squared (RMS) difference in monthly anomalies averaged 0.45 °C (ranging from 0.18 to 0.78 °C), and the RMS of station-by-station trend differences was 0.11 °C/decade. When averaged over the 97 stations, as well as into the three regional subsets, the RMS of annual anomaly differences ranged from 0.05 to 0.11 °C, and the trends were virtually identical.

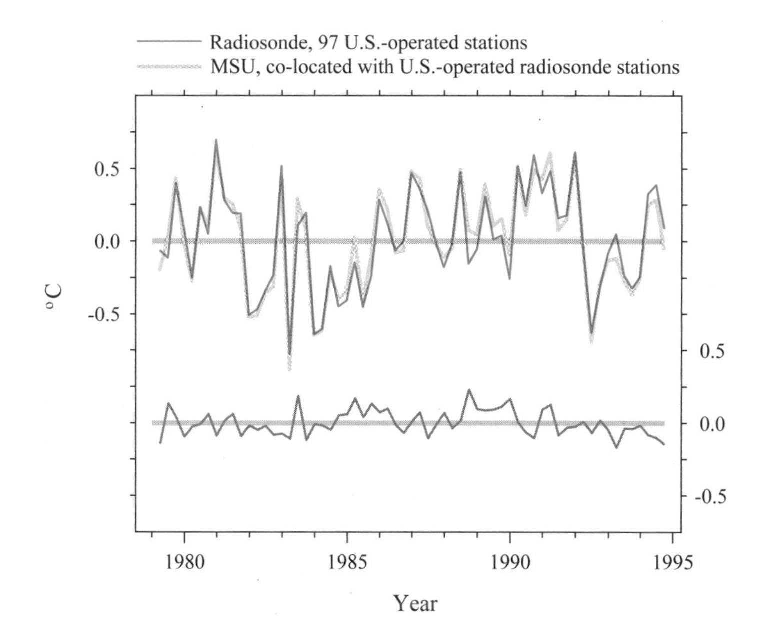

FIGURE 9.3. Time series average of radiosonde data from 97 U.S.-operated stations, believed to be free of artificial discontinuities (green curve). MSU 2LT data are from co-located grid boxes (pink curve). The dark gray curve (bottom) represents the difference between the MSU and radiosonde data sets shown in this figure. The axes are the same as in Figure 9.1, except that the data extend only through 1994.

Hurrell et al. (in review) and Santer et al. (2000) performed similar subsampling studies, but over a considerably larger geographical area. The work by Hurrell et al. compared the MSU data to a somewhat different radiosonde data set that included more radiosonde stations, but did not correct for some of the changes in radiosonde instrumentation. In the temperate latitudes of North America, the RMS of grid-point monthly anomaly differences ranged from 0.4 to 0.8 °C. In addition, Hurrell et al. examined the impact of spatial averaging methods on trends and concluded that given the spatial coherence of the troposphere, especially in the tropics, the temperature variations are mostly captured by the present sparse distribution of stations if they are latitudinally averaged. Hurrell et al. found that global trends for 1979–1998 were +0.06 (±0.11) °C/decade for MSU 2LT and +0.04 (±0.07) °C/decade for radiosondes. Santer et al. (2000) examined alternative methods of computing global-scale trends in which spatial averaging procedures and trend fitting methods were varied, giving results with larger trend differences and error bars over shorter time scales.

Among the factors that need to be considered in accounting for the differences between the results of Christy et al. (2000), Hurrell et al. (in review), and Santer et al. (2000) are: (a) differences in the size of the grid boxes over which radiosonde data are averaged before comparing them with the satellite data, (b) the different quality control criteria used for determining which stations should and should not be included in the analysis, (c) the use of daily versus monthly radiosonde reports, (d) the treatment of missing data, and (e) how the emissions from the surface that influence MSU values are handled. Further work will be required to determine the relative importance of each of these factors. In addition, given the multiplicity of decisions involved in the design of algorithms for converting satellite radiances into temperatures, the comparison of satellite and radiosonde trends should be revisited following independent verification of these data sets. Efforts to produce such data sets are already under way.

EVIDENCE CONCERNING SURFACE VERSUS TROPOSPHERIC TEMPERATURE TRENDS

Figure 2.3 shows time series of surface and tropospheric global-mean temperature anomalies. The former are based on surface station and ship observations interpolated onto a global grid, and the latter are based on the latest version (version D) of the MSU data that have been corrected for the orbital decay problems pointed out by Wentz and Schabel (1998). A difference in the trend over this 20-year period is clearly apparent. Surface temperature has been increasing at a rate of about 0.1-0.2 °C/decade, whereas tropospheric temperature has changed so little that a different sign for the trend is obtained, depending on whether or not the final year of the record is included—a year that was extraordinarily warm in the wake of the exceptionally strong 1997–98 El Niño. Spatial averages of surface and tropospheric temperature trends over the tropics/extratropics, Northern Hemisphere/Southern Hemisphere, and land/ocean exhibit qualitatively similar differences.

Direct comparison of surface and tropospheric temperature changes is feasible with radiosonde observations, because they include both surface and upper-air data. For 1979–98, Angell (1999) found the surface to have warmed more than the mid-troposphere (850 to 300 hPa layer) globally, in qualitative agreement with the results from the surface network and MSU. However, the difference was not statistically significant because of the large confidence intervals of the trends, due to the relatively short data period and small radiosonde network used. In selected high latitude regions in the Northern Hemisphere, Ross et al. (1996) found evidence of decreasing temperature trends with height from the surface through the troposphere for the period 1973–93. Using longer radiosonde data records extending back approximately forty years, Angell (1999) found less pronounced (but still noticeable) differences between surface and tropospheric temperature trends than during the satellite period, consistent with Jones's (1994) comparison of Angell's mid-tropospheric (850 to 300 hPa) radiosonde data with independent surface observations.

There is independent evidence that bears on the question of how temperature at various levels of the tropical troposphere has changed during the past 20 years. Even though the radiosonde data do not appear to show evidence of a rise in the mean freezing level in the tropics during this period (Gaffen et al., in review), tropical glaciers in the Andes and in the high mountains of Africa and Indonesia have retreated dramatically during this 20-year period (Diaz and Graham, 1996; Thompson et al., 1995; Thompson, 1999). Another indication that tropospheric temperature has increased is the fact that satellite and radiosonde measurements indicate that the water vapor loading of the tropical troposphere has increased (Wentz and Schabel, in press; Gaffen et al., 1992; Gutzler, 1992, 1996).

Although the above findings are both suggestive of a warming of the tropical troposphere, neither can be regarded as a definitive indicator of how tropospheric temperature has changed during this period. The number of tropical glaciers is quite limited, and cloudiness, precipitation, and wind speed, as well as temperature, could be factors in their mass balance. With regard to the finding that tropospheric water vapor is increasing, it should be noted that most of the water vapor loading of the tropical atmosphere tends to be concentrated within the lowest one to two kilometers of the layer sampled by the MSU measurements, and may therefore be more representative of surface temperature than tropospheric temperature.

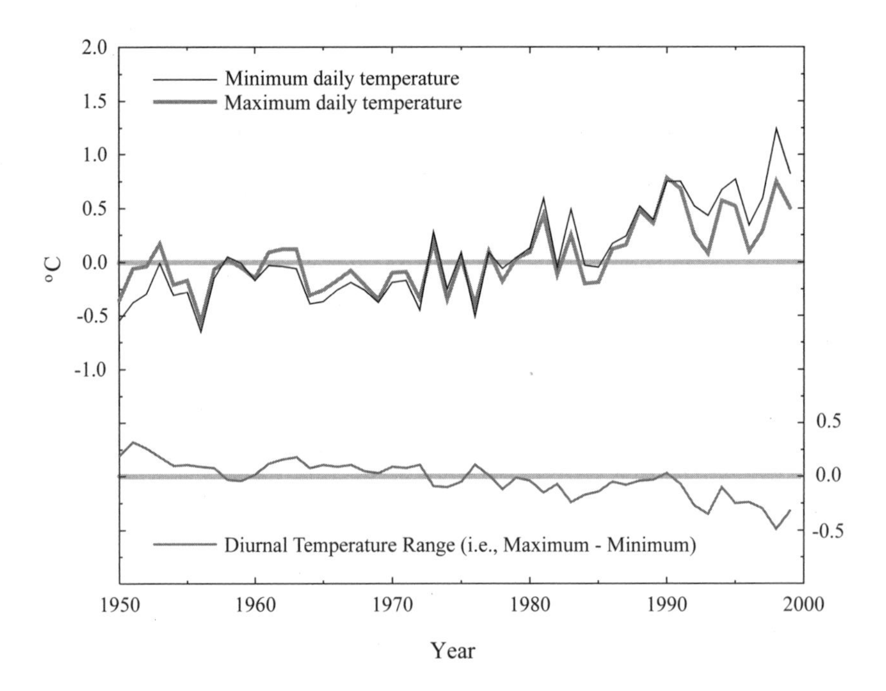

FIGURE 9.4. Time series of average daily maximum (gray curve) and minimum (black curve) land-surface air temperature anomalies, and their difference, the diurnal temperature range (maximum minus minimum; dark gray curve at the bottom). This figure, which encompasses the period 1950–99, illustrates a tendency toward a greater increase in minimum than in maximum temperature. The annual anomalies are computed as differences from the 1961–90 mean and use data from Peterson and Vose (1997).

Over many land areas, the range between daily maximum and minimum temperatures (see Figure 9.4) has been decreasing in recent decades, apparently largely in response to an increase in cloud cover (Dai et al., 1999). Daily maximum temperatures better reflect the temperature of the air mass as a whole (i.e., the tropospheric temperature), whereas the minimum temperatures often bear little relation to temperatures aloft because they are strongly influenced by the presence or absence of inversion layers close to the ground (Dai et al., 1999; Hurrell et al., in review). The fact that daily minimum temperatures have been rising at a more rapid rate than daily maximum temperatures supports the notion that surface temperature may actually be rising more rapidly than tropospheric temperature.

INTERPRETATION OF THE DIFFERENCES BETWEEN OBSERVED SURFACE AND TROPOSPHERIC TEMPERATURE TRENDS

The warming trend of approximately 0.1-0.2 °C/decade that has been observed at the earth's surface during the past 20 years clearly exceeds the observational uncertainties, including the effects of urbanization (Hansen et al. 1995; Hurrell and Trenberth, 1998). It is evident from Figure 6.2 that warming is evident in both hemispheres, at most latitudes, over most of the oceans, and over most land areas. It is also evident during all seasons of the year.

There is more of a diversity of views among panel members with respect to the degree of confidence that can be attached to the absence of a warming trend in the MSU measurements. Those more inclined to take the MSU measurements at face value cite the high degree of consistency with radiosonde measurements (Figures 2.3, 9.2, and 9.3), whereas those less inclined to do so note the retreat of the tropical glaciers and the increasing burden of water vapor (Wentz and Schabel, in press). The seasonal and interannual changes in the MSU temperature, sea surface temperature (see Figure 9.5), and atmospheric water vapor are closely coupled in the tropics according to a relatively simple thermodynamic model (Wentz and Schabel, in press). However, on decadal time scales, the trends of sea surface temperature and water vapor continue to exhibit the same close coupling, whereas the MSU and radiosonde temperature trends are less correlated. Some panel members are concerned that the satellite measurements may contain spurious discontinuities resulting from changes in the times of day (i.e., diurnal sampling) at which the satellites carrying the MSU instruments pass overhead and the influence of radiation emitted from the underlying land surface.

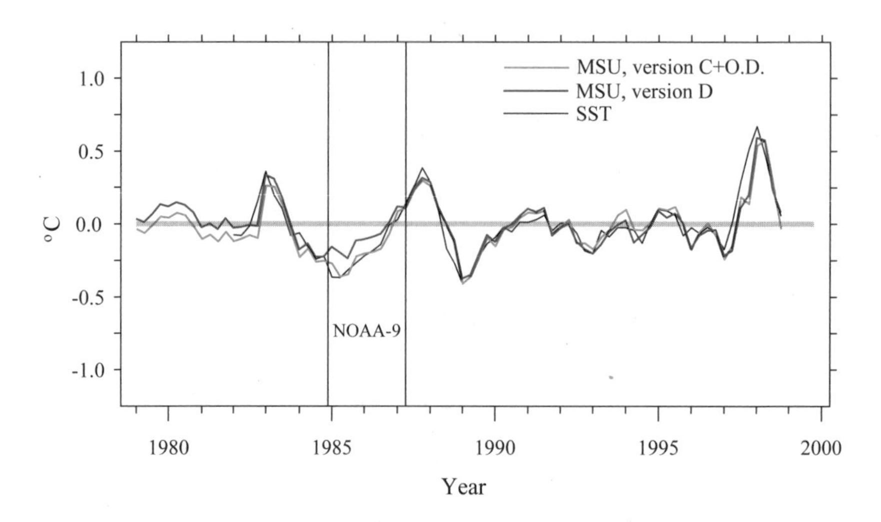

FIGURE 9.5. Lower to mid-tropospheric temperature time series from MSU version C with orbital decay corrections (C+O.D.) (aqua curve; Wentz and Schabel, 1998), version D (red curve; Christy et al., 2000), and, for reference, sea surface temperature (black curve; updated from Reynolds and Smith, 1994). The period of the NOAA-9 MSU observations is also indicated. The data are seasonal averages over the tropics (20 °N to 20 °S). The MSU data are from the period 1979–1998, and the SST data are from November 1981 through December 1998. The MSU data have been divided by 1.6 to normalize the amplitude of the tropospheric temperature variations to that of the SST data.

Some of these concerns were recently addressed with a reissuing of the MSU data set in a form in which adjustments have been made to account for orbital changes, instrument heating, and changes in diurnal sampling (see chapter 7). This adjusted version is referred to as version D. Santer et al. (in review), Christy et al. (2000), and Hurrell et al. (in review) have performed an analysis of the differences between versions C (the previous version) and D. An example of how the recent adjustments can affect the MSU temperature trend is shown in Figure 9.5. Three anomaly time series of the tropics are shown: MSU version C with the orbital decay correction (C+O.D.) of Wentz and Schabel (1998), the more recent MSU version D (Christy et al., 2000), and, for reference, the sea surface temperature data set of Reynolds and Smith (1994). A comparison of versions C+O.D. and D clearly shows a discrepancy between the two versions for the periods of the NOAA-7, -9, and -12 satellites. During the NOAA-9 period, version C+O.D. more closely tracks the SST data than version D, although version D tracks the radiosondes more closely in this same period. As mentioned above, the treatment of NOAA-9 was particularly problematic due to a relatively small inter-satellite overlap period and gain drift. In version D, a relatively large correction for gain drift is applied to NOAA-9, equivalent to 0.17 °C over two years, bringing its temperature data into closer agreement with the observations of NOAA-6 and NOAA-8, which were also observing during that period. However, the changes to the NOAA-7 and NOAA-12 data are actually greater and in the opposite direction than the adjustments to the NOAA-9 data.

Part of the observed difference between global-mean trends in surface temperature and tropospheric temperature may be a reflection of the incomplete coverage of surface data, which are sparse over the higher latitudes of the Southern Hemisphere. Recent calculations of Santer et al. (in review), based on the sub-sampling methodology described earlier, indicate that perhaps as much as one-third of the difference may be due to this effect. However, the panel views it as highly unlikely that incomplete spatial coverage of the surface data could be the primary reason for the disparity in the trends.

It seems more likely that at least part of the observed disparity is a reflection of real differences between temperature trends at the two levels. Temperatures near the earth's surface and temperatures aloft are subject to different influences, and they are often de-coupled from one another because of the presence of a temperature inversion within the atmosphere's lowest 1-2 km (Trenberth et al., 1992). Year-to-year variations in the temperature of the tropics that occur in association with El Niño tend to be about 30% greater in the middle troposphere than at the earth's surface (Hurrell and Trenberth, 1998; Wentz and Schabel, in press). In contrast, variations in circulation over the high northern latitudes exert a stronger influence on global-mean temperature at the earth's surface than in the middle troposphere (Hurrell and Trenberth, 1996). The emissions from volcanic eruptions are believed to produce stronger cooling of the troposphere than at the earth's surface (Bengtsson et al., 1999; Hansen et al., 1997).

Within a given sampling interval like the past 20-years, any changes in the temperature structure of the atmosphere that might be occurring in response to a long-term increase in atmospheric concentrations of greenhouse gases and aerosols may be masked by internal variability of the climate system (e.g., phenomena like El Niño), or variability forced by volcanic eruptions, fluctuating solar emissions, or even by short-term, naturally occurring variations in greenhouse gas concentrations themselves. For example, if the eruption of Mt. Pinatubo in 1991 were stronger and longer lasting than that of El Chichon in 1982, this would have contributed to the disparity between surface and tropospheric temperature trends of the past two decades. The longer the period of record considered, the stronger the likelihood that these naturally occurring short-term fluctuations will average out so that the observed trends are representative of the atmospheric response to longer term trends in atmospheric composition.

The modeling evidence discussed in the following section indicates that a 20-year record is subject to considerable "sampling variability" due to the presence of the short-term fluctuations in global-mean temperature discussed in the previous paragraph. As further evidence, the panel notes that according to the radiosonde record, the lower to mid troposphere warmed by a few tenths of a degree C during the late 1970s (Jones, 1994; Santer et al., 1999). Hence, the radiosonde record, to the extent that it can be believed, serves to illustrate the sensitivity of the trends to the particular choice of the period of record over which they are computed, and it suggests that the apparent lack of agreement between surface and upper air temperature trends during the past 20 years may not be representative of the longer term behavior of the climate system.

INSIGHTS DERIVED FROM MODEL SIMULATIONS

Much of our physical understanding of the climate system is encapsulated in models. Climate models are capable of simulating many of the processes that contribute to the observed differences between variations in surface temperature and tropospheric temperature, and they realistically represent the vertical structure of El Niño-related temperature fluctuations and the thermal signature of time-varying circulation patterns over higher latitudes. Unlike the climate models of a decade ago, the models used in recent simulations have enough horizontal resolution and strong enough coupling with ocean and land to realistically simulate not only the patterns of natural variability on interannual time scales, but also the amplitudes of these internal modes of variability of the climate system. Extended control runs of such models with no external forcing have been used as a basis for estimating the likelihood that the natural variability of the atmosphere-ocean system is responsible for the differing trend in surface temperature and upper air temperature (Hansen et al., 1995; Hurrell and Trenberth, 1998; Santer et al., in review). The models indicate that natural variability may indeed have contributed to the observed discrepancy, but unless the models are seriously underestimating the natural variability, it is highly unlikely that a differential trend as large as the one observed during the past 20 years could be entirely due to the internal variability of the climate system.

Models provide one way to relate observed atmospheric changes to those at the surface. This is typically done by forcing a climate model with estimates of past changes in carbon dioxide, methane, ozone, and other greenhouse gases, as well as solar variations and volcanic and anthropogenic aerosols. The types of models used to do this are either atmospheric general circulation models (GCM) coupled to oceanic GCMs, or atmospheric GCMs driven by observed SSTs. Both approaches have their relative advantages. Hansen et al. (1993, 1997), and Bengtsson et al. (1999) have explored the former approach. Bengtsson et al. note that the natural variability in globally averaged temperature time series—which typically have a standard deviation of 0.2 to 0.3 °C for a 20-year interval—makes it difficult to establish long-term temperature trends using a 20-year period. The coupled atmosphere–ocean GCM simulations of Hansen et al. (1997) suggest that dynamic coupling of the atmosphere and ocean tends to increase the variability of the troposphere-surface temperature trend difference. These studies, and those of Tett et al. (1996), highlight the importance of including the decrease in stratospheric ozone during the past 20 years, as it substantially reduces upper tropospheric warming and results in cooling in the lower stratosphere. In a series of fully coupled ocean–atmosphere and uncoupled (specified SST) climate model experiments with a range of natural and anthropogenic forcings, Hansen et al. (1997) found that different forcing mechanisms in the model simulations have different characteristic vertical temperature signatures.

Some of the forcing combinations explored by Hansen et al. (1997) yielded surface–troposphere trend differences similar to those found in the observations. Folland et al. (1998) used an atmospheric GCM forced by observed SSTs, greenhouse gases (including stratospheric and tropospheric ozone), and tropospheric sulfate aerosols for the period from 1950 to 1994, and similarly noted much improved agreement with observed temperature profile changes. They were not, however, able to reproduce the well-established strength of the observed surface warming over land. These changes have been linked in part to changes in atmospheric circulation (Hurrell, 1996), which have to be simulated to correctly reproduce the observations. In addition, it is quite possible that model–observation differences arise as a result of errors in the greenhouse gas and aerosol forcings that are applied. Thus, although climate models indicate that changes in greenhouse gases (including tropospheric and stratospheric ozone) and aerosols play a significant role in defining the vertical structure of the observed atmospheric temperature changes, model–observation discrepancies indicate that the definitive model experiments have not yet been done.

To reduce model-based discrepancies, we need better information on the changes in radiative forcings as a function of height, especially tropospheric aerosols and ozone, as well as water vapor changes and cloud changes caused by aerosols. In addition, it is likely that further work will be needed to improve model representations of the atmospheric boundary layer, clouds, and vertical heat transport in order to fully represent the observed temperature changes. Finally, models need to include more realistic representation and coupling of the stratosphere, troposphere, and ocean to fully capture the vertical structure of temperature change.

CONCLUDING REMARKS

The various kinds of evidence examined by the panel led it to conclude that the observed disparity between the surface and lower to mid-tropospheric temperature trends during this particular 20-year period is probably at least partially real. Just as the different factors that control daily maximum and minimum temperatures are evidently giving rise to a trend in the diurnal range of temperature, the different factors that control temperatures at different levels of the atmosphere are capable of altering the vertical profile of global-mean temperature. Human induced forcings (e.g., increasing concentrations of well-mixed greenhouse gases and stratospheric ozone depletion) give rise to long term changes in the vertical profile, while natural forcings such as volcanic eruptions and 'unforced' natural modes of climate variability such as El Niño give rise to large year-to-year changes which can also contribute substantially to the trends observed during a period of record as short as 20 years. The presence of these sampling variations, together with the remaining uncertainties inherent in the temperature measurements themselves, preclude the possibility of drawing more definitive conclusions concerning the cause of the observed disparity in the trends.

It is clear from the foregoing that reconciling the discrepancy between the global-mean trends in temperature is not simply a matter of deciding which one of them is correct, or determining the ideal "compromise" between them. In the long term, it will require major advances in the ability to interpret and model the subtle variations in the vertical temperature profile of the lower atmosphere that occur in association with the internal variability of the climate system in response to volcanic eruptions and solar forcing, and in connection with changes in atmospheric composition due to human activities. It will also require more precise and extensive satellite- and ground-based observations for monitoring climate change, and changes in the way these observations are implemented and processed. A detailed consideration of these issues is beyond the scope of the panel's charge (see Preface). However, the panel does offer a number of recommendations (see chapter 4) for short-term actions that it views as steps in the right direction.

References

Angell, J. K., 1988. Variations and trends in tropospheric and stratospheric global temperatures, 1958-87. *J. Climate* **1**, 1296-1313.

Angell, J. K., 1999. Comparison of surface and tropospheric trends estimated from a 63-station radiosonde network, 1958-1998. *Geophys. Res. Lett.* **26**, 2761-2764.

Angell, J. K., and J. Korshover, 1975. Estimate of the global change in tropospheric temperature between 1958 and 1973. *Mon. Wea. Rev.* **103**, 1007-1012.

Basist, A., N. C. Grody, T. C. Peterson, and C. N. Williams, 1998. Using the Special Sensor Microwave Imager to monitor land surface temperatures, wetness, and snow cover. *J. Appl. Meteorol.* **37**, 888-911.

Bengtsson, L., E. Roeckner, and M. Stendel. 1999. Why is the global warming proceeding much slower than expected? *J. Geophys. Res.* **104**, 3865-3876.

Bradley, R. S., P. M. Kelly, P. D. Jones, C. M. Goodess and H. F. Diaz, 1985. *A Climatic Data Bank for Northern Hemisphere Land Areas, 1851-1980.* TR017. Department of Energy, Washington, D.C., 335 pp.

Christy, J. R. and R. T. McNider, 1994. Satellite greenhouse warming. *Nature* **367**, 325.

Christy, J. R., R. W. Spencer, and E. Lobl, 1998. Analysis of the merging procedure for the MSU daily temperature time series. *J. Climate* **5**, 2016-2041.

Christy, J. R., R. W. Spencer, and W. D. Braswell, 2000. MSU tropospheric temperatures: Dataset construction and radiosonde comparisons. *J. Atmos. and Oc. Tech* (in press).

Conrad, V., and C. Pollak, 1950. *Methods in Climatology*, Harvard University Press, Cambridge, Mass., 459 pp.

Cryer, J. D., 1996. *Time Series Analysis*. Duxbury Press, 286 pp.

Dai, A., K. E. Trenberth, T. R. Karl, 1999. Effects of clouds, soil moisture, precipitation, and water vapor on diurnal temperature range. *J. Climate* **12**, 2451-2473.

De la Mare, W. K., 1997. Abrupt mid-twentieth-century decline in Antarctic sea-ice extent from whaling records. *Nature* **389**, 57-60.

Diaz, H. F. and N. E. Graham, 1996. Recent changes in tropical freezing heights and the role of sea surface temperature. *Nature* **383**, 152-155.

Easterling, D. R., B. Horton, P. D. Jones, T. C. Peterson, T. R. Karl, D. E. Parker, M. J. Salinger, V. Razuvaev, N. Plummer, P. Jamason, and C. K. Folland, 1997. Maximum and minimum temperature trends for the globe. *Science* **277**, 364-367.

Folland, C. K., and D. E. Parker, 1995. Correction of instrumental biases in historical sea surface temperature data. *Quart. J. Roy. Meteorol. Soc.* **121**, 319-367.

Folland, C. K., D. M. H. Sexton, D. J. Karoly, C. E. Johnson, D. P. Rowell, and D. E. Parker, 1998. Influences of anthropogenic and oceanic forcing on recent climate change. *Geophys. Res. Lett.* **25**, 353-356.

Gaffen, D. J., 1993. Historical changes in radiosonde instruments and practices. *WMO Instruments and Observing Methods Report No. 50*, WMO/TD No. 541, World Meteorological Organization, Geneva, 123 pp.

Gaffen, D. J., 1994. Temporal inhomogeneities in radiosonde temperature records. *J. Geophys. Res.* **99**, 3667-3676.

Gaffen, D. J., 1996. A digitized metadata set of global upper-air station histories, *NOAA Technical Memorandum ERL-ARL 211*, Silver Spring, MD, 38 pp.

Gaffen, D. J., W. P. Elliott, and A. Robock, 1992. Relationships between tropospheric water vapor and surface temperature as observed by radiosondes. *Geophys. Res. Lett.* **19**, 1839-1842.

Gaffen, D. J., M. A. Sargent, R. E. Habermann, and J. R. Lanzante, 2000. Sensitivity of tropospheric and stratospheric temperature trends to radiosonde data quality. *J. Climate* (in press).

Gaffen, D. J., B. D. Santer, J. S. Boyle, J. R. Christy, N. E. Graham, and R. J. Ross, in review. Multi-decadal changes in the vertical temperature structure of the tropical troposphere. *Science*.

Gullett, D. W., L. Vincent, and P. J. F. Sajecki, 1990. *Testing for Homogeneity in Temperature Time Series at Canadian Climate Stations*. CCC Report No. 90-4. Atmospheric Environment Service, Downsview, Ontario. 43 pp.

Gutzler, D., 1992. Climatic variability of temperature and humidity over the tropical western Pacific. *Geophys. Res. Lett.* **19**, 1595-1598.

Gutzler, D., 1996. Low-frequency ocean-atmosphere variability across the tropical western Pacific. *J. Atmos. Sci.* **53**, 2773-2785.

Hansen, J., and S. Lebedeff, 1987. Global trends of measured surface air temperature. *J. Geophys. Res.* **92**, 13345-13372.

Hansen, J., A. Lacis, R. Ruedy, M. Sato, and H. Wilson, 1993. How sensitive is the world's climate? *Natl. Geogr. Res. and Exploration* **9**, 142-158.

Hansen, J., H. Wilson, Mki. Sato, R. Ruedy, K. Shah, and E. Hansen, 1995. Satellite and surface temperature data at odds? *Climatic Change* **30**, 103-117.

Hansen, J., M. Sato, A. Lacis, and R. Ruedy, 1997. The missing climate forcing. *Phil. Trans. R. Soc. Lond.* **352**, 231-240.

Hansen, J., R. Ruedy, J. Glascoe, and M. Sato, 1999. GISS analysis of surface temperature change. *J. Geophys. Res* **104**, 30997-31022.

Heino, R., 1994. Climate in Finland during the Period of Meteorological Observations. *Finnish Meteorological Institute Contributions* **12,** 209 pp.

Hurrell, J. W., 1996. Influence of variations in extratropical wintertime teleconnections on Northern Hemisphere temperature. *Geophys. Res. Lett.* **23**, 665-668.

Hurrell, J. W., and K. E. Trenberth, 1996. Satellite versus surface estimates of air temperature since 1979. *J. Climate* **9**, 2222-2232.

Hurrell, J. W., and K. E. Trenberth, 1998. Difficulties in obtaining reliable temperature trends: Reconciling the surface and satellite MSU 2R trends. *J. Climate* **11**, 945-967.

Hurrell, J. W., and K. E. Trenberth, 1999. Global sea surface temperature analyses: Multiple problems and their implications for climate analysis, modeling and reanalysis. *Bull. Amer. Meteorol. Soc.* **80**, 2661-2678.

Hurrell, J. W., S. J. Brown, K. E. Trenberth, and J. R. Christy, in review. Comparison of tropospheric temperatures from radiosondes and satellites: 1979–1998. *Bull. Amer. Meteorol. Soc.*

IPCC (Intergovernmental Panel of Climate Change), 1996. *Climate Change 1995: The Science of Climate Change.* J. T. Houghton, F. G. Meira Filho, B. A. Callander, N. Harris, A. Kattenberg, and K. Maskell (eds.), Cambridge Univ. Press, Cambridge, U.K., 570 pp.

Jones, P.D., 1994. Hemispheric surface air temperature variations: A reanalysis and an update to 1993. *J. Climate* **7**, 1794-1802.

Jones, P.D., 1994. Recent warming in global temperature series. *Geophys. Res. Lett.* **21**, 1149-1152.

Jones, P. D., S. C. B. Raper, B. Santer, B. S. B Cherry, C. Goodess, P. M. Kelly, T. M. L. Wigley, R. S. Bradley, and H. F. Diaz, 1985. *A Grid Point Surface Air Temperature Data Set for the Northern Hemisphere.* TRO22, Department of Energy, Washington, D.C. 251 pp.

Jones, P. D., T. J. Osborn, and K. R. Briffa, 1997. Estimating sampling errors in large-scale temperature averages. *J. Climate* **10,** 2548-2568.

Jones, P. D., M. New, D. E. Parker, S. Martin, and I. G. Rigor, 1999. Surface air temperature and its changes over the past 150 years. *Rev. Geophys.* **37**, 173-200.

Kalnay, E., M. Kanamitsu, R. Kistler, W. Collins, D. Deaven, L. Gandin, M. Iredell, S. Saha, G. White, J. Woollen, Y. Zhu, M. Chelliah, W. Ebisuzaki, W. Higgins, J. Janowiak, K. C. Mo, C. Ropelewski, J. Wang, A. Leetmaa, R. Reynolds, Roy Jenne, and Dennis Joseph, 1996. The NMC/NCAR 40-Year Reanalysis Project. *Bull. Amer. Meteorol. Soc.* **77**, 437-471.

Karl, T. R., and C. N. Williams, Jr., 1987. An approach to adjusting climatological time series for discontinuous inhomogeneities. *J. Climate Appl. Meteor.* **26**, 1744-1763.

Karl, T. R., P. D. Jones, R. W. Knight, G. Kukla, N. Plummer, V. Razuvaev, K. P. Gallo, J. Lindseay, and T. C. Peterson, 1993. A new perspective on recent global warming: Asymmetric trends of daily maximum and minimum temperatures. *Bull. Amer. Meteorol. Soc.* **14**, 1007-1023.

Karl, T. R., R. W. Knight, and J. R. Christy, 1994. Global and hemispheric temperature trends: Uncertainties related to inadequate spatial sampling. *J. Climate* **7**, 1144-1163.

Luers, J. K., and R. E. Eskridge. 1998. Use of radiosonde temperature data in climate studies. *J. Climate* **11**, 1002-1019.

Mo, T., 1995. A study of the Microwave Sounding Unit on the NOAA-12 satellite. *IEEE Trans. Geoscience and Remote Sensing* **33**, 1141-1152.

NRC (National Research Council), 1999. *Adequacy of Climate Observing Systems*. National Academy Press, Washington, D.C., 51 pp.

Oerlemans, J., 1994. Quantifying global warming from the retreat of glaciers. *Science* **264**, 243-245.

Oort, A. H., and H. Liu, 1993. Upper-air temperature trends over the globe, 1958-1989. *J. Climate* **6**, 292-307.

Owen, T. W., K. P. Gallo, C. D. Elvidge, and K. E. Baugh, 1998. Using DMSP-OLS light frequency data to categorize urban environments associated with U.S. climate observing stations. *Intl. J. Remote Sensing* **19**, 3451-3456.

Parker, D. E., and D. I. Cox. 1995. Towards a consistent global climatological rawinsonde data-base. *Internat. J. Climatol.* **15**, 473-496.

Parker, D. E., P. D. Jones, A. Bevan, and C. K. Folland, 1994. Interdecadal changes of surface temperature since the late 19th century. *J. Geophys. Res.* **99**, 14373-14399.

Parker, D. E., M. Gordon, D. P. N. Cullum, D. M. H. Sexton, C. K. Folland, and N. Rayner, 1997. A new global gridded radiosonde temperature data base and recent temperature trends. *Geophys. Res. Lett.* **24**, 1499-1502.

Parkinson, C. L., D. J. Cavalieri, P. Gloersen, H. J. Zwally, and J. C. Comiso, 1999. Arctic sea ice extents, areas, and trends, 1978-1996. *J. Geophys. Res.* **104**, 20837-20856.

Peterson, T., H. Daan, and P. Jones, 1997. Initial Selection of a GCOS Surface Network, *Bull. Amer. Meteorol. Soc.* **78**, 2145-2152.

Peterson, T. C., and J. F. Griffiths, 1997. Historical African data. *Bull. Amer. Meteorol. Soc.* **78**, 2869-2872.

Peterson, T. C., and R. S. Vose, 1997. An overview of the Global Historical Climatology Network temperature database. *Bull. Amer. Meteorol. Soc.* **78**, 2837-2849.

Peterson, T. C., T. R. Karl, P. F. Jamason, R. Knight, and D. R. Easterling, 1998a. The first difference method: Maximizing station density for the calculation of long-term global temperature change. *J. Geophys. Res.* **103**, 25967-25974.

Peterson, T. C., D. R. Easterling, T. R. Karl, P. Ya. Groisman, N. Nicholls, N. Plummer, S. Torok, I. Auer, R. Boehm, D. Gullett, L. Vincent, R. Heino, H. Tuomenvirta, O. Mestre, T. Szentimre, J. Salinger, E. Førland, I. Hanssen-Bauer, H. Alexandersson, P. Jones, D. Parker, 1998b. Homogeneity adjustments of in situ atmospheric climate data: A review. *Int. J. Climatol.* **18**, 1493-1517.

Peterson, T. C., R. S. Vose, R. Schmoyer, and V. Razuvaev, 1998c. GHCN quality control of monthly temperature data. *Internat. J. Climatol.* **18**, 1169-1179.

Peterson, T. C., K. P. Gallo, J. Lawrimore, T. W. Owen, A. Huang, and D. A. McKittrick, 1999. Global rural temperature trends. *Geophys. Res. Lett.* **26**, 329-332.

Pollack, H. N., S. Huang, P. Y. Shen, 1998. Climate change record in subsurface temperatures: a global perspective. *Science* **282**, 279-281.

Prabhakara, C., R. Iacovazzi, Jr., J.-M. Yoo, G. Dalu, 1998. Global warming deduced from MSU. *Geophys. Res. Lett.* **25**, 1927-1930.

Quayle, R. G., T. C. Peterson, A. N. Basist, and C. S. Godfrey, 1999. An operational near real time global temperature index. *Geophys. Res. Lett.* **26**, 333-336.

Reynolds, R. W., and D. C. Marsico, 1993. An improved real time global SST analysis. *J. Climate* **6**, 114-119.

Reynolds, R. W., and T. M. Smith, 1994. Improved global sea surface temperature analyses using optimum interpolation. *J. Climate* **7**, 929-948.

Reynolds, R. W., and T. M. Smith, 1994. Improved global sea surface temperature analyses using optimum interpolation. *J. Climate* **7**, 929-948.

Ross, R. J., J. Otterman, D. Starr, W. Elliott, J. Angell, and J. Susskind, 1996. Regional trends of surface and tropospheric temperature and evening-morning temperature difference in northern latitudes: 1973-93. *Geophys. Res. Lett.* **23**, 3179-3182.

Rothrock, D. A., Y. Yu, G. A. Maykut, 1999. Thinning of the Arctic sea-ice cover. *Geophys. Res. Lett.* **26**, 3469-3472.

Santer, B. D., J. J. Hnilo, T. M. L. Wigley, J. S. Boyle, C. Doutriaux, M. Fiorino, D. E. Parker, and K. E. Taylor, 1999. Uncertainties in observationally based estimates of temperature change in the free atmosphere. *J. Geophys. Res.* **104**, 6305-6338.

Santer, B. D., T. M. L. Wigley, J. S. Boyle, D. Gaffen, J. J. Hnilo, D. Nychka, D. E. Parker, and K. E. Taylor, 2000. Statistical significance of trend differences in layer-average temperature time series. *J. Geophys. Res.* (in press).

Santer, B. D., T. M. L. Wigley, D. J. Gaffen, L. Bengtsson, C. Doutriaux, J. S. Boyle, M. Esch, J. J. Hnilo, P. D. Jones, G. A. Meehl, E. Roeckner, K. E. Taylor and M. F. Wehner, in review. Interpreting differential temperature trends at the surface and in the lower troposphere. *Science.*

Smith, T. M., R. E. Livezey, and S. S. Shen, 1998. An improved method for analyzing sparse and irregularly distributed SST data on a regular grid: The tropical Pacific Ocean. *J. Climate* **11**, 1717-1729.

Spencer, R. W., and J.R. Christy, 1992. Precision and radiosonde validation of satellite gridpoint temperature anomalies. Part II: A tropospheric retrieval and trends during 1979-90. *J. Clim.* **5**, 858-866.

Tett, S. F. B., J. F. B. Mitchell, D. E. Parker, and M. R. Allen, 1996. Human influence on the atmospheric vertical temperature structure: detection and observations. *Science* **274**, 1170-1173.

Thompson, L.G., E. Mosley-Thompson, M.E. Davis, P.-N. Lin, K.A. Henderson, J. Cole-Dai, J.F. Bolzan, and K.-B. Liu, 1995. Late Glacial Stage and Holocene tropical ice core records from Huascarán, Peru. *Science* **269**, 46-50.

Thompson, L.G., 1999. Ice core evidence for climate change in the Tropics: Implications for our future. *Quaternary Science Review* (in press).

Trenberth, K. E., J. R. Christy, and J. W. Hurrell, 1992. Monitoring global monthly mean surface temperatures. *J. Climate* **5**, 1405-1423.

Uppala, S., 1997. Observing system performance in ERA. *ECMWF Re-Analysis Project Report Series*, No. 3. European Centre for Medium-Range Weather Forecast, Reading, U.K., 261 pp.

Vinnikov, A., A. Robock, R. J. Stouffer, J. Walsh, C. L. Parkinson, D. J. Cavalieri, J. F. B. Mitchell, D. Garrett, V. F. Zakharov, 1999. Global Warming and Northern Hemisphere Sea Ice Extent. *Science* **286**, 1934-1937.

Wallis, T. W. R., 1998. A subset of Core Stations from the Comprehensive Aerological Reference Dataset (CARDS). *J. Climate* **11**, 272-282.

Warrick, R. A., C. Le Provost, M. F. Meier, J. Oerlemans, and P. L. Woodworth, 1996. Changes in sea level, *Climate Change 1995: The Science of Climate Change.* Intergovernmental Panel on Climate Change, Cambridge University Press, 359-407.

Wentz, F. J., and M. Schabel, 1998. Effects of orbital decay on satellite-derived lower tropospheric temperature trends. *Nature* **394**, 661-664.

Wentz, F. J., and M. Schabel, in press. Precise climate monitoring using complementary satellite data sets. *Nature.*

WMO (World Meteorological Organization), 1994. *Guide to the Applications of Marine Climatology.* WMO - No. 781. 119 pp.

WMO, 1996. Measurements of upper air temperature, pressure, and humidity. *Guide to Meteorological Instruments and Methods of Observation*, chapter 12. WMO-No. 8, sixth edition, Geneva, I.12-1–I.12-32.

Woodruff, S. D., H. F. Diaz, J. D. Elms, and S. J. Worley, 1998. COADS release 2 data and metadata enhancements for improvements of marine surface flux fields. *Phys. Chem. Earth* **23**, 517-526.

Appendixes

Appendix A

Biographical Information on Panel Members

John M. Wallace (Chair) is professor of atmospheric sciences and co-director of the University of Washington Program on the Environment. From 1981-98 he served as director of the (University of Washington/NOAA) Joint Institute for the Study of the Atmosphere and Ocean. His research specialties include the study of atmospheric general circulation, El Niño, and global climate. He is a member of the National Academy of Sciences; a fellow of the American Association for the Advancement of Science, the American Geophysical Union (AGU), and the American Meteorological Society (AMS); and a recipient of the Rossby medal (AMS) and Revelle medal (AGU).

John R. Christy is professor of atmospheric science at the University of Alabama in Huntsville. He specializes in satellite microwave data in evaluating global climate change. He has appeared as an expert at congressional hearings and is a member of NASA's Global Hydrology and Climate Center which focuses on climate research. He was recently named by the American Meteorological Society to receive a special award, jointly with Dr. Roy W. Spencer of the Marshall Space Flight Center, for developing a global, precise record of earth's temperature from operational polar-orbiting satellites which is regarded as having advanced scientists' ability to monitor climate. Data from the Spencer-Christy research project is used in both national and international policy analyses relating to global climate change and for validating climate models.

Dian Gaffen leads the climate variability and trends group at the NOAA Air Resources Laboratory in Silver Spring, Maryland. Her recent research focuses on observational studies of atmospheric temperature and water vapor changes, climate extremes, and meteorological data quality. She is a member of the American Meteorological Society and the American Geophysical Union and a recipient of both the Prof. Dr. Vilho Vaisala Award from the World Meteorological Organization and the NOAA Administrator's Award.

Norman C. Grody is affiliated with NOAA NESDIS (National Environmental Satellite Data and Information Service) where he has developed techniques to retrieve atmospheric parameters (e.g., temperature, water vapor, rainfall) and identify surface features (e.g., snowcover, sea ice, flooding) using satellite-based microwave radiometers. He has received the U.S. Department of Commerce Bronze and Silver Medal Awards for the development of operational products from the SSM/I and AMSU instruments, respectively.

James E. Hansen is head of the NASA/Goddard Institute for Space Studies. His research interests include radiative transfer in planetary atmospheres, interpretation of remote sounding of planetary atmospheres, development of simplified climate models and three-dimensional global climate models, current climate trends from observational data, and projections of man's impact on climate. He is a member of the National Academy of Sciences and a fellow of the American Geophysical Union.

David E. Parker is with the Hadley Centre for Climate Prediction and Research at The Meteorological Office in the United Kingdom. Since 1979 his work has focused on climatic variability and change and on near-real-time monitoring of climatic variations. He has contributed to the development of global historical data bases for sea surface temperature and sea ice, as well as marine air temperature, mean sea level pressure, and radiosonde-based air temperatures with a view to the detection and attribution of climate changes and the forcing and verification of climate model simulations. He is a fellow of the Royal Meteorological Society; a contributor to the 1990, 1992, 1995, and current IPCC Assessments; and a recipient of the Fitzroy Prize of the Royal Meteorological Society.

Thomas C. Peterson is chief of the Scientific Services Division at the NOAA National Climatic Data Center. His expertise lies in assessing surface data and surface climate variability and change, including analyses of various temperature characteristics. He is a WMO CCI rapporteur on statistical methods for climatology and serves as chair of the Joint WMO CCI/CLIVAR Working Group on Climate Change Detection. He has received the U.S. Department of

Commerce Bronze Medal Award "for developing revolutionary new climatological baseline data sets and statistical techniques that reveal accurate long-term climatic trends." He is a member of the American Geophysical Union and the American Meteorological Society.

Benjamin D. Santer is a physicist/atmospheric scientist at the Program for Climate Model Diagnosis and Intercomparison at Lawrence Livermore National Laboratory. His research interests include identifying human effects on climate and evaluating the performance of global climate models. He is the recipient of a MacArthur Fellowship, the 1998 Norbert Gerbier-MUMM International Award from the World Meteorological Organization, and the 1997 NOAA/Environmental Research Laboratories "Outstanding Scientific Paper" award.

Roy W. Spencer is Senior Scientist for Climate Studies at NASA's Marshall Space Flight Center. His research has focused on satellite information retrieval techniques, passive microwave remote sensing, satellite precipitation retrieval, global temperature monitoring, space sensor definition, and satellite meteorology. He is a recipient of NASA's Exceptional Scientific Achievement Medal and a co-recipient, along with Dr. John Christy, of the American Meteorological Society's Special Award for their global temperature monitoring work with satellites. Dr. Spencer is the U.S. Science Team Leader for the Advanced Microwave Scanning Radiometer that will fly on NASA's Aqua spacecraft in 2000. He is a member of the American Meteorological Society.

Kevin Trenberth is head of the Climate Analysis Section at the National Center for Atmospheric Research (NCAR). His main scientific interests are in global climate variability and its effects, including El Niño and global climate change, and being from New Zealand, Southern Hemisphere meteorology. He is currently a member of the NRC's Committee on Global Change Research, the NOAA Advisory Panel on Climate and Global Change and Council on Long-term Monitoring, the Joint Scientific Committee of the WCRP, the National Science Foundation's Climate System Modeling Advisory Board, and the ECMWF Reanalysis Project Advisory Group. He is co-chair of the International Scientific Steering Group (SSG) for the World Climate Research Programme (WCRP) Climate Variability and Predictability (CLIVAR) Programme and chair of the Center for Ocean-Land-Atmosphere Studies Scientific Advisory Committee. He has also been prominent in the IPCC Scientific Assessment activities. He is a Fellow of the American Meteorological Society and the American Association for the Advancement of Science and an Honorary Fellow of the New Zealand Royal Society.

Frank J. Wentz established and currently serves as director of Remote Sensing Systems, a research company specializing in satellite microwave remote sensing of the earth. His research focuses on radiative transfer models that relate satellite observations to geophysical parameters, with the objective of providing reliable geophysical data sets to the earth science community. He is currently working on satellite-derived decadal time series of atmospheric moisture and temperature, the measurement of sea-surface temperature through clouds, and advanced microwave sensor designs for climatological studies. He is a member of the American Geophysical Union.

Appendix B

Acronyms and Abbreviations

BASC	Board on Atmospheric Sciences and Climate
COADS	Comprehensive Ocean/Atmosphere Data Set
CRC	Climate Research Committee
ENSO	El Niño–Southern Oscillation
GCM	general circulation model
GHz	GigaHertz
hPa	hectoPascal
IPCC	Intergovernmental Panel on Climate Change
MSU	Microwave Sounding Unit
MSU 2LT	lower to mid-tropospheric temperature derived from MSU channel 2
NOAA	National Oceanic and Atmospheric Administration
NRC	National Research Council
RMS	root-mean-square
SST	sea surface temperature
UKMO	United Kingdom Meteorological Office
UTC	Coordinated Universal Time
WMO	World Meteorological Organization